IT WASN'T MEANT TO BE LIKE THIS

IT WASN'T MEANT TO BE LIKE THIS

The unintended but entirely predictable consequences of keeping sheep...

ANDY OFFER

MOSAÏQUEPRESS

Published by
MOSAÏQUE PRESS
Registered office:
Bank Gallery
High Street
Kenilworth, Warwickshire CV8 1LY

Cover photo by kind permission of Claire Whiteman-Haywood

ISBN 978-1-906852-76-4

*'We never intended to farm for real.
We had a few hobby sheep at our remote
Herefordshire home and sold lamb to friends
and family. Sometimes we wished
it had stayed like that, but it didn't.
Welcome to the story of Whyle House Lamb.'*

CONTENTS

Part 1: LEARNING TO FARM

Prologue   8

1. How we got here   13
2. The hobby years   26
3. Farming by accident   55
4. Could do better   88
5. Learning the hard way   124

Part 2: FARMING FOR REAL

6. Finding our own way   153
7. Moving up a gear   180
8. We've made it!   202
9. Where next?   215
10. Back to a normal life   233

Acknowledgements   241

PROLOGUE

A LAMB AND HER MUM peered out hopefully from one of our makeshift pens. It wasn't feeding time yet so they'd have to wait but there was something amiss. I was sure this pen had twins in it that morning, I remembered thinking how well they looked.

I consulted our 'book', a once-pristine notebook labelled 'Lambing 2014', its blank pages so full of promise and expectation – now dog-eared, stained with unmentionable fluids and barely legible, with entries in several different hands. 'Pen 25, Ewe 478, Twins,' I eventually deciphered.

I called Frances, who was working next door. 'Who's this in pen 25?'

'Oh I moved her and put a single in there this morning,' came the reply. 'Didn't I tell you?'

Now, we'd learned two things about lambing over the years.

The first was that its mostly about record-keeping and being organised rather than the obstetric dramas you see in the 'How to Lamb' books. The other, which is far more important, is that working with your other half, twenty-four hours a day, when you're both exhausted and the end is still far out of sight, is not a recipe for marital bliss. Lambing sheds can be tense places, especially when things are going wrong and the simplest misunderstanding can lead to conflagrations on an epic scale.

We always laughed with hindsight at my 'lambing shed temperament', a phrase my daughter, Hannah, coined after a particularly incendiary incident, but it could be tough at the time. The aim was always reasonably peaceful coexistence.

So in this case, it was clearly my fault. Like most men, I listen avidly to everything my wife tells me but on this occasion I'd clearly forgotten. I summoned all my reserves of tact and patience and through only marginally gritted teeth replied, 'No worries, I'll change the records. Have you got the tag reader out there?'

'"Pen 25, Ewe 478, Twins," I eventually deciphered.'

I heard a muffled shout and a crash. I decided to leave it for a few minutes before continuing.

Trying to do my bit for marital harmony and as penance for my forgetfulness, I went to collect the reader rather than ask her to bring it to me.

That reader had changed our lives. Our sheep were all identified by tiny electronic codes held within the tags in their ears. 'A bit like bar-coding for sheep,' as grandson Ollie had once described it. No more crawling around with a torch, trying to read faded, grimy so-called 'visual' tags. We just pointed it at a sheep's ear and her number appeared on the screen! And we could then record other information like pen numbers, number of lambs and any treatments we used. On this occasion, I just wanted to record the new pen occupants and then go and find the family that Frances had moved earlier.

I pointed the reader at the ewe's ear. Nothing. I tried again. Still no reassuring beep. Maybe it was the tag. I tried the ewe in the next pen. Still nothing. I began frantically pressing buttons to try to wake it up. Still nothing.

Frances was eying me cautiously. 'It's not broken is it? I might have put it down a bit hard.'

'A bit hard?'

'You made me jump when you called and I dropped it on the floor. It hit the metal gate on the way down. So it's at least partly your fault!'

With an enormous effort, I smiled. Eight hundred quid down the drain but she looked so remorseful. And the last thing I broke was a tractor and that cost several thousand pounds to fix so I wasn't on especially high moral ground. 'Never mind,' I muttered, 'these things happen.'

An uneasy silence prevailed as I updated the manual records. 'Let's go back to the house for a coffee,' I suggested. 'We've both had enough.'

'Anybody there?' came a man's voice from outside. We looked at each other uncertainly. Partly because visitors during lambing are unusual and partly because we were wondering how much of our somewhat heated discussion he'd heard.

Frances opened the door and with a huge, and very forced smile greeted Cecil, our farming neighbour.

'Hi both,' he said cheerily, 'I've got a proposition for you.'

We looked at each other, Frances's face was still flushed and I was sure mine was too.

'I was asking Bert the other day,' Cecil continued, 'about bidding for his arable ground when he packs up but I didn't really want the grass. He told me he knew a man who did! So I was wondering if you'd like to put in a joint bid with me for the whole farm? I wouldn't presume to know your financial position but it occurred to me that you might not be able to make the numbers work for the whole acreage.'

I looked at Frances who eyed me cautiously but with the shadow of a smile. I was tempted to jump out of the pen I was standing in and shake his hand there and then but decided to play it cool.

'Cecil, we're really chuffed to be asked,' I began, 'and I reckon it would be a great opportunity, but can we think about it overnight?'

'Yeah, of course,' he replied. 'You've a lot to be thinking about.'

'We need to look at some numbers and see if they add up.'

'In farming they rarely do,' he laughed, 'but when you've decided, pop down and we'll have a drink on it. I'll get out of your way now. It looks like you've plenty to do!' He winked at Frances with a smile.

And with that he was gone.

I was finding it hard to contain my excitement; maybe this was our big break? A proper farm. Right next door! We'd worked so hard to get to this point but we were struggling to move the business on. With this land we'd save time, a fortune on fuel driving round all our rented fields and I'd be able to do most of the farm work on my own with the dog. Frances could have a bit more time to herself. And we'd have room for lots more sheep. We'd be real farmers at last!

But the best bit was, we'd been asked. Cecil was a respected farmer and was now confident enough to offer us this joint bid. No longer the 'comers-in with income off the farm, just playing at it, who'll probably disappear as quickly as they arrived.' We were part of the community and were accepted.

We'd come a long way on our farming journey but we needed to get bigger. Maybe this was the chance we'd been waiting for.

1. How We Got Here

I'VE ALWAYS WANTED to farm. As far back as I can remember, I've been fascinated. In the early days it was the machines – like all small children I loved watching them working and I remember peering through the hedge to watch the new combine, more properly called a combine harvester, working in the field next door. Tiny by modern standards, it was so slow I could hardly see it moving, but even at that age I was riveted. We were high up on the Chiltern hills in north Hertfordshire, on the edge of a tiny village between Hitchin and Luton, barely thirty-five miles from central London and less than twenty-five miles from the M25 – or at least where the M25 would be some thirty years later. But this was another world, it was 1959 and I was just four years old.

By the time I reached nine or ten, I was able to help. Long before health and safety got in the way and when farming was more manual, there were many jobs my friend Neil and I could help with. The men on the neighbouring farm were encouraging and let us do things that wouldn't be allowed nowadays. We minded the grain store when they were in the fields harvesting – we weren't big enough to fix things but were deemed capable of turning it off if it went wrong. And to underline how important we were, we even got paid – the handsome sum of £3-10 shillings (£3.50) a week. And that was for nine-hour days, seven days a week, which would probably involve a visit from the authorities nowadays. They also let us drive the tractor between the heaps of straw bales when they were loading the trailers to cart them back to the farm yard. Strictly illegal, well out of the boss's sight and enormous fun!

But despite my enthusiasm, getting into farming wasn't going to be that simple. My family weren't farmers, I had no farm to inherit nor the money to buy one. So if I wanted to be involved, I was destined to be an employee and thereby came the second problem. The family were wary, even discouraging. Mum saw farming through the eyes of her village friends – a low-paid, hard, physical job. Dad had a thing about taking all your opportunities and as I was doing well at school, he thought I should aim for a professional job.

And school didn't help either. It was one of the unexpected challenges of going to a grammar school. Decades before league tables and Ofsted reports, our school was a good one with a reputation for good results, good behaviour and great sporting achievement. I like to think I managed two out of three but this farming thing was becoming wearying. If I mentioned it in class I'd get laughter from the other boys with cries of 'Old Farmer Offerrrr' and indulgent smirks from the teachers. They just didn't get it.

My careers interview comprised a single question: 'Farming's a bit risky with the weather, isn't it?' 'Blimey, I hadn't thought of that,' I muttered to myself.

A HOT CRACKLY AUGUST afternoon in 1969 saw me sitting in a tractor in Frank's field, waiting for his combine to fill. He was our other farming neighbour and I was actually quite nervous now: this was for real. At fourteen I wasn't a kid any more; this was my first proper farm job and I didn't want to mess it up. Frank's son Hugh was on the combine – a lovely chap who taught me a great deal over the years but not the most patient with 'corn cart' drivers. Especially those who held him up so that he had to stop harvesting.

I was watching him like a hawk. There were no cabs on combines in those days and eventually I saw him wave. I started the tractor and immediately stalled it as I pulled away. I could see Hugh watching me. I started it again and drove towards the combine. I pulled alongside and stopped a few yards ahead so I could engage the right gear. I glanced nervously across at Hugh who gave me a thumbs-up.

Does that mean he's ready or 'Are you ready?' I wondered. It meant both.

He pulled a lever and the corn cascaded into the trailer. I lifted the clutch and moved tentatively forward, trying to keep the corn falling in the middle. I struggled to watch where I was going while keeping the grain from falling outside the trailer.

The first tankful unloaded without mishap and I drew away and waited for him to wave again.

This time, I pulled alongside, engaged the right gear but pulled away too slowly. The grain shot over the front of the trailer before Hugh had time to turn it off. He glared at me and pointed up the field, so I moved further along and started again. This time we managed to get it all in the trailer, much to my relief.

My farming bosses didn't share the family's or school's doubts about my choice of career. They encouraged me towards agricultural college, their reasoning being that I'd need proper qualifications if I was to work towards a farm manager's job. I liked this idea – I'd still be farming but with someone else's money.

But school intervened again. I was doing A-levels by this time with eyes on the local college, but they insisted I was university material and that I should do a degree.

So I applied for and was offered a place at Newcastle University to study agriculture. The compromise, the first of many as it turned out, was that I'd do a degree but would still aim to move into farm management.

AT THAT POINT I had a bit of luck. Newcastle insisted that I had a year on a farm before joining the course – what we'd now call a gap year. The logic was that I wasn't a farmer's son, so needed a whole year of practical experience. Ridiculous as it turned out as most of my contemporaries were farmer's sons and daughters who had only experienced their parents' farm, whereas I had worked on several different ones and knew a little about most systems.

But it was a welcome opportunity to have a rest from studying and to earn some money – and this was where I met Ian, possibly the most influential guide and mentor I had during my learning years. I worked for him for twelve months, learned that farming doesn't just happen in the school holidays and there are periods when there's little going on. I learned about dairy farming, how to rear young stock, and how to work as part of a young, energetic and hardworking team.

Dad's unease about my career choice was relieved by the fact that I was going to university but Mum was still unhappy. She worried about the long hours and the fact that I often came home exhausted. I'm sure she was only concerned for my welfare but it was hard work trying to convince her.

My main achievement during that year off was learning how to look after and milk cows. The regular cowman had left and Ian had temporarily taken over and taught me to milk as his 'relief'. Sadly, at that point he became gravely ill and I had to take over full-time for a while.

IT WAS 4.15 AM as the alarm went off. I was wide awake anyway, knowing I'd got my first morning milking on my own. I dragged myself out of bed, wiped a flannel over my face and stumbled downstairs. The light was on in the kitchen and my Mum, bless her, had got up to make my flask.

I walked quietly out into the still-dark morning and jumped into my old van. Luckily it started first time and I made my way through the lanes to Ian's farm. Driving at this time of day was strange. There were no lights on anywhere, all the houses and cottages were in darkness and I saw nobody on the road. I arrived in the yard at about twenty to five. Too early really – I could have a few more minutes in bed next time. It was just getting light as I walked down the track to fetch the cows and there was a cacophony of noise coming from the woods – my first dawn chorus.

They were waiting for me, the boss cows at the front. They pushed the gate open as I released it and made their way, calmly and unhurriedly up to the yard. I followed them, making sure I'd got them all. I'd missed some one afternoon when it was foggy and had to go back for them, much to Ian's amusement.

The cows stood quietly in the collecting yard while I turned on the milking parlour and connected up the pipes and valves.

My early milking career was defined by a series of 'Whatever you do, don't....' warnings.

17

'Don't leave the pipe unconnected and spray milk all over the floor.'

'Don't forget to turn the water heater on and fill up the cold water tank during milking or you'll have to wait for ages at the end to wash down.'

'ALWAYS, always, double-check the valve for the wash cycle at the end or you'll put cleaning agent in the tank and the whole lot will have to go down the drain.' This was folklore on that farm as a previous employee had actually done this and caused a whole day's milk to be wasted.

And finally, 'Don't leave any cows behind in the field!'

It took about two hours and was a deeply relaxing almost cerebral process, driven entirely by routine. The same cows came in pretty much in turn each time, the boss ones first and the timid ones last. Panda was always last and I had to go out to fetch her. She got her name via a rather convoluted association between her number, PC49, a legendary 1960s TV policeman and the small patrol cars they used, nicknamed 'Panda Cars'.

As I was washing down, Horace, our tractor driver, peered round the parlour door. 'Just checking to make sure you got enough milk for your cornflakes,' he laughed. In his eyes I'd now gone over to the dark side and become a cowman – not to be trusted and definitely inferior to the rest of the team.

SOME WEEKS LATER, the milk recorder, a pleasant and efficient woman named Helen, was due. She came to two milkings, one afternoon and one morning, took a sample from each cow and recorded her yield. The essential prerequisite for this was that I could recognise each cow. Which I couldn't. I knew the boss ones and the timid ones and Panda and her Mum who was still in the herd and just as daft, but I couldn't recognise them all as Ian could. The solution we came up with was for Ian, recently out of hospital, to come and sit in the parlour as I milked in the afternoon and tell me who each cow was.

We got on fine: Helen took her little pot of milk and recorded the yield. We then put a 'tail tape' on each cow with her number written on in permanent black marker, so I could identify them in the morning. It was a slow job, adding no more than a minute to each cow, but with eighty of them, that was an extra hour. But it would all be worth it in the morning.

I arrived early the next morning, about four-thirty, so I could get the cows in and get organised before Helen arrived. As I walked up the track behind the cows listening to the now familiar dawn chorus and marvelling at how peaceful the world was at this time, I suddenly noticed that one of the tail tapes was dangling from a tail. I walked up behind her and stuck it back on, but as I did so, I noticed another and another and one or two where it'd completely gone. My stomach tightened; this was going to be a tough one. It had rained in the night and these tail tapes were clearly not rain fast.

I decided to try and wing it, but Helen noticed as soon as the first cows came in.

'It's OK, I know these ones,' I said breezily. She looked unconvinced but we muddled through a few batches. Some I knew and some still had their tapes on. And then eventually, we were faced with a row of completely anonymous cow backsides and tails.

'Can we just ignore these?' I pleaded.

'Not really, or the records will be incomplete. I've got the ear numbers here as well as their herd numbers. Can't we use those?'

'Well we can, but it means scrambling up to the front of each cow to read the tattoos in their ears and the light isn't very good in here.'

'Don't worry, I have a torch in the car!'

I shrugged my drooping shoulders. This was going to be a very long job and my breakfast seemed a very long way off.

I eventually sat down at the table nearly two hours late and Ian was amazed that I'd struggled through reading ear numbers. 'How on earth did you see them?'

'With great difficulty,' I muttered.

'I'd have told her to get lost.'

'Yes but you're the boss,' I countered. He smiled. 'Well done anyway, I think we can claim that you're a real cowman now you've survived a milk recording.'

'Don't say that to Horace!' I pleaded.

In my last week with Ian I learned to drive a combine which had a pleasing symmetry about it. The four-year-old kid peeping through the fence, longing to be part of farm life and watching that tiny combine go up and down the field in a cloud of dust was now, fifteen years later, actually driving one. A fitting end to a great year and I was now ready to go off to Newcastle and learn how to do it all properly.

WHILE AT UNIVERSITY, I was encouraged into my second compromise. My tutor was adamant that I shouldn't go into farm management. 'You'll just be the employee that they don't pay overtime,' was his argument. Plus the fact that all the good management jobs on the large estates were filled from the old boys' network and I wouldn't get a look-in.

So I applied for and was offered a consultancy job with ADAS, the advisory service for farmers, which at the time was run by the government.

On results day I called home. 'I got a 2:1,' I said excitedly.

'That's nice,' said Mum.

'And I was top student,' I added. 'I'm now all set to join ADAS – and get into farming properly!'

'But it's really just gardening on a bigger scale,' she replied,

'I don't see why you needed all this studying.' I put the phone down and smiled weakly. I'd long since given up trying to win them over.

'I bet your parents are thrilled,' said a friend as he picked up the phone to make the same call.

'Delighted,' I said. And I'm sure, in their own way, they were really.

I took the ADAS job and embarked on a thirty-year career in farm advisory work. The relevance to this story is the experience I gained, adding flesh and substance to my degree skeleton, year by year, season by season. Seeing what worked on the better-run farms in my patch, helping others to do the same. 'Sharing best practice,' we called it. As I moved up through the business, I learned how to manage staff and money and after we were privatised, how to keep customers happy and the board content.

I'd concluded by this time that I was never going to be able to farm in my own right. It was a distant ambition, fading away as my work and family responsibilities increased. I'd met Frances by then and although we both hankered after a more rural home, we had no real ambition beyond a nice place to live and maybe a few acres for me to play with.

So perhaps the most significant contribution ADAS made to this story was that it gave me the income, the financial wherewithal, to buy Whyle House.

IT WAS AN EVENTFUL and convoluted journey to finding our new home. The owners were as bizarre as some of the properties we viewed. One back in 1999 that we very nearly bought got us unwittingly into an unpleasant neighbours' dispute.

The bedside phone was ringing at seven thirty on a Sunday morning. 'Jarvis here, I can't make two this afternoon, you'll have to come at three.' Not so much a question, more of a command.

'OK,' I said, 'I'll see you then, I want to talk about...' But he'd rung off.

As instructed, at three o'clock I made my way to the property, fifteen acres next to the river Avon, just outside Pershore in Worcestershire. The house was an old timbered dwelling that the agent described as 'needing updating' but that in the real world needed a major renovation. Our plan was to pay for this work by selling off some small strips of land to neighbours to extend their gardens. The first three we'd talked to had been fine and were interested, but Mr Jarvis would be our immediate neighbour and the others had warned us that he was 'difficult'.

Difficult proved to be an understatement. He launched into a tirade as soon as I walked through the door.

'What do you think you're going to do with that place? There's no way I'm paying a penny over agricultural value, and I know what I'm talking about.' The implication was that I didn't.

'The house is falling down and water comes up through the floor. The whole place floods, you'll never grow anything there.' He was working himself up into a rage when the penny dropped. He wanted the place himself.

We found out later that he believed he'd had first refusal but the owner was in severe financial difficulties and the bank had instructed the agent to sell. The agent had taken a dislike to Mr Jarvis (I can't think why) and accepted our offer instead. The following weekend there was a heavy rain storm and we decided to go and check on the flooding around the property. The fields had flooded although the house looked secure. We saw Mr Jarvis peering out through his curtains at us as we drove down the lane. 'I'm not sure I want this hassle,' I said.

'And the flood risk is real,' agreed Frances.

On the following Monday I called the agent and withdrew our offer.

We looked at another property a few weeks later, down a long muddy track. An elderly man with unkempt hair and trousers held up with baler twine hobbled across the yard to meet us and cast his arms around.

'Well this is it,' he said with the air of someone who was once proud of what he had but had lost interest. He'd also lost a good deal of cash by the looks of it. 'These are grand cattle yards. We've had some good beasts through these.'

'Mmm, ' I mumbled, eyeing the loose tin cladding clattering in the breeze and the holes in the roof.

'Have a look round and come to the house when you're done. Don't worry about your boots; we don't.'

I looked down at his mud-encrusted footwear. This was not looking promising.

We had a cursory look round. The place was a mud bath, battered tin sheds, collapsing brick walls. Nothing had been touched for years, maybe decades. The house was no better. We did take our muddy shoes off as we stepped inside. The elderly man was sitting in a filthy old chair next to a woman who we assumed was his wife. Another man, a little younger but just as down at heel, sat on an old stool with his back against a wooden table, eyeing us cautiously.

'Go an' have a look round dears,' said the woman. 'We can't get upstairs that easy now.'

We groped our way up the musty stairs; the mice tracks between the wallpaper and the walls were a bit disconcerting. Everywhere there was the smell of neglect, of damp and of poverty. We'd seen enough. 'It's a lovely spot,' we enthused, 'But it needs quite a bit of time and money spent and we've got neither, I'm afraid.'

'Don't worry dears,' said the woman, 'I don't 'spect you'd like having to start the generator every morning either, would you?'

'We viewed Whyle House for the first time on a beautiful late summer morning in 2001 and fell in love at first sight.'

'What generator?' I asked incredulously.

'We've no mains 'lectric here,' muttered the old man. 'But it's no problem – it starts pretty good.'

It was hard to believe – less than twenty miles from Worcester, surrounded by modern property but the place had no mains power. A place that time just forgot.

As we turned to leave, the younger man leaning on the table suddenly spoke. 'I know where I've seen you two before. You were going to buy my place down by the river!' We found out later that his financial affairs were such that he couldn't afford heating and spent his time with this couple to keep warm.

WHYLE HOUSE HAD COME onto our radar some months before we viewed it but it was marketed as a fully renovated equestrian property, so we'd not pursued it. When it came through a second time, I decided to call the agent, the delightfully named Margot, to seek more information.

I explained that we didn't have horses and that we were really hoping to put our own mark on a property rather than buy one fully renovated. 'Well it's not been renovated much or that well,' she responded, trying a little too hard not to lose the sale.

We viewed Whyle House for the first time on a beautiful late summer morning in 2001 and fell in love at first sight. We had a long list of 'must haves' with absolute essentials. One absolute was that it must be near the motorway network as we were both still working.

Whyle House was nearly an hour from the M5. 'But we came a complicated way,' we observed. 'We're sure there's a quicker way.' There wasn't of course and there were lots of other 'must haves' that we ignored. But it was what we wanted and we agreed as we drove away that we'd make an offer. We were at the end of the terrible period of foot-and-mouth disease and the vendors were desperate to sell as they'd already bought another house, so we did the deal over the autumn and moved in just before Christmas.

The vendors may have been desperate to sell but this sense of urgency didn't extend actually to moving out. We arrived as the light was failing on a December Friday afternoon with two lorry loads of furniture to find them both still packing. We eventually got them out of the house and garage that evening but it was Easter the following year before they were finally gone from the barns and sheds.

The next day I had a message from my ex-mother-in-law. It's safe to say that we'd never been on the best of terms, especially since her daughter and I had long gone our separate ways, but the sentiment was genuine. 'We're so pleased you've managed to realise your childhood dream and bought a farm. Good luck and we hope it works out for you.' It wasn't really a farm, as we shall see. But it turned out to be a huge first step on the long journey to achieving that dream.

2. THE HOBBY YEARS

'AND WHERE ARE YOU OFF TO?' called Frances, brandishing her moving-in to-do list. It was a grey morning, mizzly, not cold but damp and uninviting, as I opened the door and for the first time looked out onto our new farm.

'I'm going to walk around our land,' I said rather loftily. 'You coming with me?'

She came, although she didn't really see what the urgency was. We tramped up to the top of our fields where the agent had shown us the views of Hay Bluff. But of course we couldn't see it that day in the mist. We paddled through the mud where our farming neighbour Bert's cattle had churned up the gateways and made our way across to the pond. It was full of water but overgrown with trees and bushes.

'I'll need to clear that out and fence it,' I said excitedly. 'In fact the whole place needs re-fencing. And look at that old shed! We can replace that with a better building at some stage.'

'Let's get the house sorted first, shall we,' said Frances, the would-be (or perhaps would rather not be) farmer's wife.

We made our way back to the house, through Bert's cattle, happily grazing our lower fields. Frances was thinking about curtains and furniture and I was revelling in the fact that I'd finally become a farmer. It'd been my dream, my ambition forever, and never in my wildest moments did I ever think I'd be able to do it. But here I was with my very own eight acres. A postage stamp by commercial farming standards but it was mine and I was beyond happy.

A few days later I was chatting with a family member. 'I hear you've bought a small holding', she said brightly.

'No,' I growled, 'it's a small farm. Small holdings are for people who eat nuts and wear dungarees.'

We wanted to farm commercially and we wanted to develop a profitable and robust business but we just didn't have the resources to do it in a big way at the outset. So, my more polite response to the family member should have been, 'No, we're going to farm commercially but in a small way.'

I spent many happy hours walking our tiny fields and debating how we might farm them. I'd done this so often in my ADAS job, helping farmers organise their businesses and I was having a ball. I had a friend who was an agricultural engineer and he'd painstakingly put together a set of equipment to grow wheat on a few acres at home. I thought that sounded like fun but it really wasn't practical. We couldn't possibly afford to invest in the machinery needed to grow crops and we could never persuade a contractor to cultivate and harvest them on that scale – even if we had somewhere to store the grain – which of course we didn't.

And that wasn't the only reality check I needed in those euphoric early days. I realised very quickly that I was viewing my new holding through a 1970s lens; I was no longer that young, or that fit. Those wonderful days with Frank and Ian had been fun and exciting – I was working with a team and I'd had no financial responsibility. But it was just the two of us now and it was our own money we were spending. We were both still working which limited our available time, so our plan had to be constrained by the available money, time and frankly, energy.

IT WAS OBVIOUS REALLY that grass farming with animals would be the easiest option. The machinery requirements were minimal and once we'd got the place properly fenced we could start slowly with a few animals and see how we got on. So after much debate, we left everything exactly as it was and kept our grass fields.

The next step was to decide what animals we would have to graze them. I wanted some cattle. I'd worked with them before, and I knew how to handle them but again, it wasn't really practical. Our land was too wet to keep them outside in the winter and we had no suitable housing for them. Our holding was also in one of the UK's hot spots for bovine TB and although it wasn't part of our thinking at the time, I've reflected frequently since that our decision not to have cattle was the right one for that reason alone.

So we decided to keep sheep. Probably the most significant decision we ever took and one which shaped our entire future business.

There was another reason for choosing sheep over cattle, a much more personal one. Most of my help was willing but inexperienced. They generally made up for lack of knowledge with limitless energy and endless, if somewhat unrestrained, enthusiasm. But cattle are big and heavy and can be danger-ous. They need handling by skilled people who know how to

'So we decided to keep sheep.'

keep themselves safe and the animals calm. All animals can be impulsive but a sheep deciding to do something unexpected might result in a few bruises or some damaged pride, whereas with cattle the results can be very serious if not fatal.

We had a brief flirtation with organic farming at this stage. I'd often wondered about it and I was interested to see if it would work on our small acreage.

I spent some time looking into it and concluded pretty quickly that it wasn't for us. Talking to organic farmers confirmed our fears. We'd have to plough up all our grassland and sow clover as we couldn't use nitrogen fertiliser. Our husbandry would need to be faultless as we'd have limited access to drugs if the stock got sick. On top of that, the number of sheep we could keep per acre (the stocking rate) on a decent organic system was so low that we'd barely get into double figures on our acreage.

The other issue was the cost of the organic registration. We'd need it to get the premium prices, but it would be out of all proportion to our turnover.

It was really a problem of scale in those early days, with the lower output per acre and accreditation fees making an organic system uneconomic. I wasn't afraid of hard work and I recognised that sheep farming wasn't very profitable at the best of times, but even I wouldn't set out deliberately to lose money!

In truth we weren't that committed ourselves and even today organic farming has a degree of idealism about it. It's a very good marketing brand – but that's all it is. There is little or no evidence that it is better for consumers or for the environment and on many farms, especially where it's used as an excuse not to do things, it's most definitely not better for the stock.

HAVING DECIDED TO KEEP sheep, we needed to agree what type and where to get them from. The choices were overwhelming – what breed would suit us, should we buy cheap older sheep or expensive young ones, and should I risk buying them in the market or find some locally? I decided to ask Sue, a former colleague and friend who farmed near Worcester. 'Why don't you buy some ewe (female) lambs,' was her advice. 'You could keep them for a year to see how you get on and then decide whether to tup (breed from) them or sell them on as yearling ewes.'

I don't think she really believed we'd stick at it but her suggestion seemed sensible and we'd nothing to lose. So I had a word with Bert and he took me up the field in his Land Rover to see some of his ewe lambs. 'Have a look through them and choose which ones you want. They'll be £70 each.' There was a tone in his voice that suggested he wouldn't haggle and I wanted to keep him on side in case I needed more help, so I agreed and suggested he pick me ten good ones and drop them off the following week.

On 12th September 2004 we got our first ten Suffolk cross Mule ewe lambs and I proudly put them in our orchard. My first proper farm animals!

'We kept these lambs outside through the winter...'

They settled in quickly although they were wild and very easy to spook. I reasoned this was because they'd only been handled by Bert with his dogs and decided that they'd benefit from some human handling. We got them penned up and I noticed one was limping. After a struggle, I managed to catch it and found it had a maggoted foot. In a panic we called Bert who hacked at the foot with his penknife before dousing it with sheep dip. 'A sheep's worst enemy is maggots!' he shouted as he turned her over onto her three good feet. 'She'll be fine – if not I'll replace her.'

She did get better and maggots are indeed a sheep's worst enemy but there are now much better ways to treat them. And I don't think we had a maggoted foot problem after that in all our years of farming.

We kept these lambs outside through the winter and I walked up the field before work each morning with hay and lamb pellets for them. They grew well and really blossomed on the new grass in the spring. I enjoyed having them around; they were a good antidote to the day job and despite Sue's doubts I was getting hooked. Even Frances found them in-

31

teresting and we decided to keep them and produce our own lambs the following year.

I realised we'd stepped onto the farming conveyor belt, or was it a treadmill? These few sheep needed protecting against worms and flies and it became clear very quickly that we needed some proper handling facilities to gather and catch them. We also needed equipment to apply the treatments and so the list went on. My first project was a treatment race or corral made from corrugated iron I found in the shed and an old garden gate which had followed us through several house-moves as being something which will 'come in handy' one day.

My endless quest to save money by trying to make my own equipment is a recurring theme. Hay racks, feeding barriers in the barn, a handling yoke as well as the treatment race were all fashioned from bits of wood and other recycled materials, but inevitably these home-spun items had to be replaced eventually with proper commercial kit. We wasted a lot of time over the years, messing about with inadequate equipment.

AS NEW SHEEP FARMERS, we needed to make sure we kept them healthy and this is how we met Harriet the vet. I'd known her name for years as she used to work with ADAS on various animal health trials and demonstrations but I had no idea that she was working at our local practice. This was amazingly lucky as she was a nationally known sheep veterinary specialist who became a great friend.

Her approach created an immediate rapport. She was as keen on stopping things going wrong as she was on diagnosis and treatment. She was also the partner of a local sheep farming friend and so had first-hand practical experience as well as a wealth of knowledge. She had huge credibility with local sheep-keepers and was very busy.

I got to know her well over the years. She always arrived

with a ready smile, usually late as she was in so much demand, and always did the job with humour, very efficiently and with great tolerance of my shortcomings – especially in the early days. 'Are you sure you want to do that' was her way of gently nudging me in the right direction.

Unlike many sheep farmers who see calling the vet as the last resort, or even as a sign of failure, I built a relationship with Harriet and used her as my mentor and consultant as well as my vet. She was interested in the broader aspects of sheep management including nutrition and we spent many hours over the years looking at our flock performance data and trying to improve our productivity and profit. We had good-natured disagreements of course, mostly about the condition of my ewes and how I fed them and she did have the habit of finding more problems than I realised I had. I used to joke that our role was to provide sick animals for her to photograph for use in her talks to farmers' meetings, but despite this we valued her vast experience and knowledge and she remained a vital part of our team throughout our farming journey.

In those early days, she sat me down and told me a few home truths about farmers and vets.

'You've done the right thing by getting me out before you have a crisis,' was the first one. 'If you've shown me your stock, sought my advice and made it clear that you want a long-term relationship with me, then, when you call me on Sunday afternoon or late at night, I'll know you're not wasting my time. I'll also know where you live!'

Another one was, 'Make sure you have a decent gathering/catching pen or race and are willing to catch and restrain the animals to help me. If I arrive and the animals are still in the field and, even worse, you expect me to catch them, then I'll be a lot less willing to come out next time!'

And finally, 'I won't – indeed I'm not allowed to – prescribe medicines for you unless your stock are 'known to me'. There will be times when you'll need an extra antibiotic spray or a bottle of anti-inflammatory drug and if I know the health status of your flock I'll be much more likely to prescribe it for you to collect from the surgery, which is more convenient for you and me.'

WE WERE MAKING PROGRESS. We had our first ten sheep and had acquired some rudimentary equipment to handle them. We'd built a good relationship with Harriet but I still had nagging doubts about my own ability. I was a 'corn and cows' man – I really didn't know much about sheep. So I decided I needed some training.

I left work early one day as I'd enrolled on a Sheep Farming for Smallholders course – despite my reservations about 'small holdings' I consoled myself that at least it was at a recognised agricultural college. A bit like the first day at school, we were standing self-consciously in the corridor outside the lecture room, wondering what credentials we all had for being on the course. And then Charlotte arrived, a local sheep farmer's wife and Head of Agriculture at the college. She was a small bundle of energy and enthusiasm, great fun and knew her subject thoroughly. We started with Sheep Breeds and worked our way through all the theory of nutrition and sheep care. As Charlotte farmed herself, she could sprinkle pieces of hard-won practical knowledge into her lectures which made it all very real and just a bit challenging. We learned that sheep can get over sixty different diseases, which caused a few groans, and she told us that she'd had people drop out at this point. By way of reassurance, she explained that they don't get all these ailments at once. Not especially reassuring, admittedly, but we carried on.

I decided to remain incognito and not declare my farming background so that I was treated just like all the others. My fellow trainees ranged from a retired vicar, keen to minister to a different flock, as he put it, to some rather idealistic young smallholders. There was also an elderly couple who wanted sheep in their orchard as lawn mowers. They'd just been conned by someone charging them £60 to pregnancy-scan two ewes (the going rate was about fifty pence per animal then) and as a result they'd decided they needed some knowledge.

The practical days on this course were held on Saturday mornings at the farm that Charlotte ran with her husband, Oliver. I arrived on the first morning with muddy wellies – a schoolboy error on a livestock farm. I was very relieved that they didn't know my ADAS background as I sheepishly washed them off with the hose.

I nearly blew my cover on the first practical day as I was more confident than the others with the animals. After I'd crutched a ewe (removed the soiled wool from its backend) using the electric clippers, Charlotte gave me a quizzical look. 'You've done this before, haven't you,' she said.

'No, honestly, I haven't,' I replied, which was the truth but I don't think she was convinced.

All the sessions were run with a smile and great tolerance by Charlotte and Oliver. At one of the sessions, I gently enquired what a typical price would be for ewe lambs. I'd just bought more from Bert and was concerned I'd paid too much for them. Without hesitation, Charlotte said, 'The price you pay in the market is public knowledge; what you pay in a private deal stays private.' So that was that.

I can honestly say that I enjoyed every minute of that course. I learned to turn sheep over – an essential skill as it immobilises the animal so that you can inspect/treat or do whatever else you need to do with it. I also practised dosing and drenching, injecting, foot-trimming, crutching and how to assess body condition.

The highlight though, was the lambing day at Charlotte's farm. They were lambing six hundred or so ewes with just the two of them and, given how busy he was, Oliver was a fantastic teacher. I had with me Celine, my French daughter-in-law who Charlotte insisted on calling my '2IC' – second in command – and we got some real hands-on practice at spotting signs of lambing and dealing with the difficult ones.

During our coffee break, Charlotte and Oliver finally asked me outright what my background was and I came clean. They both laughed. 'We just knew you weren't as green as you pretended to be.'

I learned two critical things on that lambing day. The first was the value of routine so that things don't get forgotten when you're tired or when crises occur, and the second was that you mustn't let your French daughter-in-law spray a continental '7' on an English-born sheep!

'How the hell am I going to sell that in Hereford! You'd better hope it wears off before it goes!' was Oliver's parting shot.

THE FINAL SESSION on this course, later in the year, was on shearing sheep. It was a lovely warm morning as we assembled in the shearing shed at the college. The smell of sheep and lanolin pervaded the air and was worn into the polished wooden shearing floors. Our teacher was a kindly older man who told us he'd been shearing all of his life and that there was nothing too it really – it was just practice. He showed us how to set up the 'handpiece' or shearing clippers which fitted onto a long flexible drive from the motor.

'Set the comb forward for safety and back for speed,' was his advice.

'Safety sounds good,' I suggested.

'No, it's safety for the sheep,' he grinned. That was me told.

He opened the sheep pen door and grabbed his first ewe,

gently turning her over onto her bottom as he did so. I was confident I could do that after the earlier course sessions, but he made it look so easy.

He then proceeded to remove her wool and she sat there quietly while he did so. Not a wriggle or a struggle, just a nice calm experience with neither man nor beast breaking sweat. He tugged the rope to stop the shears, released the ewe who bounded away, feeling much lighter and more comfortable, and grabbed the fleece, which amazingly was all in one piece. He then showed us how to roll the fleece with the inside facing out and how to make a wool rope with the tail end to secure the roll.

I wasn't sure whether to be inspired or defeated before we started. Ewes are heavy animals and I was sure they wouldn't behave as well as that for me.

I walked cautiously over to the shearing 'stand' and picked up the handpiece – it was very heavy and I could feel my arm aching before I even started. I pulled the rope to start it and it nearly twisted out of my hand. I stopped it quickly, placed it back on the floor and opened the pen door. I tried to find a small ewe but they were all pretty big. I grabbed one and tried to turn it over as we'd been taught. Unsurprisingly, it didn't want to cooperate and we both staggered backwards around the pen until eventually it lost its balance (just before I was about to) and I got it turned onto its bottom.

'Don't balance it on its tail bone,' yelled the teacher, 'it hurts them and they'll wriggle.'

I got the ewe settled and reached down for the handpiece. The ewe saw an opportunity to escape as I bent over and I had to drop the handpiece and re-settle her. I tried again and this time managed to stand upright with the ewe in position and the handpiece in my left hand – the wrong hand. I was sweating before I even started to shear. This was going to be a long day.

Eventually I parted my first sheep from its wool and as she bounded off to rejoin her friends, I laid the fleece out to roll.

'That's not bad for a first time,' said the teacher encouragingly. 'It's even nearly all in one piece!'

They say pride comes before a fall and I proved that with my next sheep. 'I'm pretty good at this,' I thought as I grabbed the second one. But it was even bigger than the first and much less cooperative. It wriggled and jumped and I cut it with the shears, which didn't help our rapport. With sweat running into my eyes, I finished it. The sheep broke free and rejoined the others. Instead of a whole fleece, I was left with a pile of loose wool which I despondently tried, unsuccessfully, to roll.

'Do one more and then we'll have a break,' said the teacher, 'You've only ten to do and there's plenty of time.'

I resisted the urge to walk away, telling myself, 'I didn't have to do this. It was only for interest and so I knew how it was done.' But I stayed and completed my third sheep with reasonable competence.

During our break, the teacher showed us how to hand-shear. It was even more relaxed than powered shearing and totally silent apart from the gentle click of the clippers. The sheep almost fell asleep as he gently removed her wool.

'I always like to hand-shear a few each year, just to keep my hand in,' he said reassuringly.

We murmured our approval as we wandered over to the chairs and slumped down for a rest and a drink.

I completed three more sheep before lunch but I was getting tired and my arms and shoulders ached. So that was six done in a morning – professional shearers can do twenty times that number before lunch and almost as many again in the afternoon. I was really beginning to hurt now. What I thought was a lovely summer morning had turned into a sweaty, smelly and downright painful experience.

After lunch I returned reluctantly to my stand and grabbed my next sheep. I felt a bit better after our rest and had a little more control over the animal but my energy was flagging fast and I was struggling. She sensed this and decided to make a break. She kicked hard onto the floor and as my grip loosened for a moment, she wriggled and broke free, I dropped the handpiece which was still chattering away as it hit the floor, I silenced it with the stop rope and ran off after the sheep. She trotted insolently across the shearing floor, dragging a half shorn fleece in her wake. I eventually cornered and caught her but I couldn't turn her over. I'd no strength left. The teacher saw me struggling and came to help.

'I'm sorry, I can't do any more,' I whimpered pathetically. 'Can you finish her?'

He smiled, grabbed her and effortlessly turned her over so he could drag her back to the stand. I did at least manage to roll her fleece and claimed that I'd sheared my first six and a half sheep.

Charlotte's final bit of wisdom as we left was, 'You should treat shearing a bit like bricklaying: once you know how to do it, you realise it's a job for someone else!'

This course was brilliant but sadly and probably inevitably it has been discontinued. It didn't earn much for the college and Charlotte has now retired.

Apart from the practical knowhow, I gained some much needed perspective from that course...

Things don't generally go wrong. Mostly sheep try hard to stay well and can look after themselves.

The only way to learn to lamb is to do it – with someone who knows what they're doing

And just because you can do the hard jobs like shearing, doesn't mean you should.

'It was with some anxiety that I called Adrian.'

OUR NEXT CHALLENGE was to get our own ten sheep shorn – and with Charlotte's words ringing in my ears, I decided to find a contractor to do it.

It was with some anxiety that I called Adrian. Most professional shearers won't get out of bed to shear ten ewes but I couldn't do them myself. His croaky, heavily Welsh accented voice on the phone suggested someone in their sixties so maybe he was retiring shortly and was happy to take on small jobs? We arranged for him to come the following Sunday afternoon. I organised some help and started to plan the penning arrangements.

Sunday afternoon was hot and sunny and we got our 'flock' in to keep them out of the sun and to make sure we were ready. Adrian arrived in his battered little car. He was a slight eighteen-year-old lad who didn't look solid enough to handle sheep. A farmer's son from Pen y Bont in Wales and fresh out of college, he was taking on small jobs to build his experience and reputation. That he'd been to college at all marked him out as a bit different around here, and his good nature and slow easy smile appealed to us immediately. His first job was unhurriedly to reorganise my pens and the shearing area so that it would work in practice.

He couldn't have been anything but Welsh with his looks and he was slender but goodness he was strong. Turning over sheep is largely a matter of technique as I have found to the cost of my back over the years, but he just had the knack of someone born to it and he sheared our ten ewes in no time. Having Frances and me plus a couple of friends was probably overkill as a squad to shear ten animals but he just smiled as we tripped over each other, trying to manoeuvre the sheep into position for him. He even patiently showed us how to roll the fleeces properly as I'd learned on the course but since forgotten. He charged us the going rate for shearing (about £1 a ewe then) plus his petrol money, which was beyond reasonable.

He was a truly lovely guy who became a friend and we would look forward to seeing him every year. Even when we were shearing two hundred ewes, or when I dragged him out to shear ewe lambs in September, he still retained that slow easy smile and unhurried nature.

A lesson I learned very early from Adrian is that if you treat contractors with courtesy and consideration, they'll do right by you. He liked coming to us because we were organised. We'd book several weeks ahead and if the weather was bad, we'd always get the sheep in well beforehand so they were dry. We had good handling facilities and plenty of help so he knew he could do a full day's work. It's amazing how many farmers would cancel on the day 'because the sheep were wet' or worse still, call him and expect him there that day as 'they had a bit of time spare.'

NOW THAT WE FELT more proficient as sheep-keepers and were gaining confidence, it was time to think about our first lambing, planned for the following spring.

'If you want to lamb those ewes of yours, you'll need a tup,' a statement that was biologically undeniable. 'Would you like to borrow one of mine?' offered Sue, our trusty friend in those early days. 'I'll need him first but if you can wait until November, you can borrow Basil. He's a young Texel and I've had some good lambs out of him.'

I realised this would save us a lot of money and accepted her offer gratefully. We planned to lamb at Easter when we both had some holiday so Basil arriving late in the season worked perfectly – the old, 'Tup in on bonfire night gives you an April fool' rule or thereabouts.

In mid-November 2005, Basil arrived for what Sue called 'his holidays' and set about getting our animals pregnant. He had a crayon or Raddle on a harness round his neck to mark the ewe's backsides when he mounted them so we could see which animals had been served. We were keen to do it right, so we changed the colour of the raddle each week during the three-week service period, so we could work out which animals would lamb in which week.

Charlotte had instilled in us the need to get the ewes pregnancy-scanned but like shearing, it's difficult to find someone who will scan just ten sheep. Eventually we found John who lived locally and who agreed to call in one evening on his way home.

'It's just like human scanning except the ewes don't go around boring everyone rigid with photographs,' he quipped as he set up his equipment in a hastily made pen in the corner of our field. It was getting dark but he was methodical and calm. The sheep ran through his trailer which had a shoot or race attached and he held them gently with a foot-operated

'In mid-November 2005 Basil arrived for what Sue called 'his holidays'.'

restraint round their shoulders while he applied his scanning machine to their undersides.

'Twins,' he said quietly. John did everything quietly which was a great help with nervous first-time mums (and shepherds). We marked her neck with some green marker spray.

'Twins,' again – more green spray.

'Single.' That was a blow: I'd hoped, with unrealistic optimism, that we might get a full two hundred percent with everyone having twins. Orange spray this time.

'Empty.' That couldn't be right. But it was. 'Do you want to separate her so you can get rid of her?'

'No let her go – we'll sort her later.' It hadn't crossed my mind that we might have barren or empty ewes.

'You ought to mark her or you won't be able to distinguish her from ones where the paint has worn off.' Wisdom that we applied in later years but too late this time as we'd let her go with the others. But we'd keep her anyway – she ought to have a second chance.

We continued with the rest and John gave me a little slip

*'Our first-ever lamb was born on 10th April 2006 –
a single ram lamb to ewe number 654!'*

of paper with my first-ever scanning percentage – one hundred and fifty percent – six sets of twins and three singles from ten ewes.

'Having an empty with such a small number of ewes knocks the percentage down heavily,' said John trying to reassure us. 'At least you've no trebles.'

I was a bit deflated, I was hoping for more than that, but I couldn't do anything about it. That was our maximum limit set for our first lambing. All we had to do now was keep them all alive.

John charged us the going rate per sheep which was fifty pence and I gave him a tenner. It took him longer to disinfect his gear than it did to do the job. Like Adrian the shearer, we made a good friend of John and we used him every year after that.

Looking back it's laughable really. All this work for ten sheep but we really did want to do things properly. I built some pens in the shed so we could separate the singles (and the barren one) from the twins and feed the twins more and restrict the singles. This was so their lambs didn't get too big and give us problems at lambing. Keeping them inside helped us because we were still working at this stage and it made it easier for us to feed them before and after work when it was dark outside.

Our first-ever lamb was born on 10th April 2006 – a single ram lamb to ewe number 654! We'd been watching them for days, checking every hour or so but now it was arriving, I panicked. I couldn't pull it out and was about to call for help when it plopped out of its own accord. That happened several times during our first lambing and I sent Frances off each time to fetch neighbour Bert to help me, only to find that the lambs arrived naturally a few minutes later. Afterwards, I discovered that Frances never actually went to fetch him – she just walked around outside for a few minutes and then came back in when all was peace and quiet again.

During our first lambing we lost one lamb – it was born backwards and suffocated – and that was a case of lack of experience on my part. We tried to revive it by swinging it round by its back legs as taught on the course, but we were inexperienced and forgot that we should dry its legs or hold them with a towel. It slipped through my hands and hit the wall with a sickening thud. 'Probably just as well it was already dead,' muttered Frances.

The other notable birth during that first lambing was Minimo. On Good Friday 2006 this tiny ewe lamb came into the world – her brother had had all the food in the womb and she was so small she could sit in your hand. My daughter-in-law Celine – the one who sprayed a French 7 on Charlotte's lamb – kept her alive against all the odds, even buying her a red plastic 'lambmac' which had to be cut down to fit her. Of course, the problem here was that nobody wanted to see Minimo go to the abattoir and she was too small to breed from. They say, 'Never give a lamb a name,' for a reason.

'...a red plastic 'lambmac' cut down to fit her.'

Fifteen lambings and several thousand lambs later, I can confirm that by and large, ewes know what they are doing. In the later years we probably assisted only two or three ewes out of two hundred or so and lambs lost to late or non-intervention averaged less than one per year.

FARMING MEAT ANIMALS means that their eventual demise is an inevitable fact of life. There's no getting away from it: if we were going to make any money at this farming game, then our animals had to be killed. But it didn't make it any easier and I'd been uncomfortable about this for some time. We'd not given them names (except Minimo) but with only fourteen animals, it was hard not to get to know them. We went through the lambs and selected two that were ready to go. A task that we'd eventually do every Monday morning but on this occasion it was a tough call.

I loaded them into my home-made trailer and we set off for the abattoir, having done a timed dummy run the day before. We had a slot when we were supposed to arrive (we found out later that everyone was given the same time slot) and we didn't want to be late. I drove slowly and Frances and I sat in silence.

It didn't seem appropriate to put the radio on somehow. We arrived on time and backed up to the lairage gates. This was a task in itself as the yard is long and narrow and backing a small trailer the full length of it wasn't easy. We got the lambs unloaded, having searched to find someone to show us where to put them. We were then asked for a cutting list which I didn't have so I wrote one out for each lamb against an ear tag number.

'We don't use tags here,' came the terse reply. 'We'll just cut them up as they come. If you want them done differently you'll have to mark them with spray.' Suitably chastened, we left the lambs to their fate and returned home.

Our first abattoir visit wasn't a comfortable one and customer care is clearly not high on their list of priorities. But over the years we got to know the staff and they're actually a great bunch of lads. I guess good communication skills and client engagement are not top of the list for someone working in that environment day after day.

The day we were due to collect the lambs from the abattoir didn't go quite as planned. My old university friend and his wife called to invite us out for lunch. Bill and Helen lived in Cumbria so we didn't see them very often and we were keen to catch up. I explained our plans and Bill's response was, 'Don't worry, we'll come with you. We can stop for lunch on the way!' It was lovely to see them and we had a great chat over an extended lunch. We then drove over to the abattoir to collect the lambs and were presented with two plastic bags.

'We've done a standard cut for you as we didn't have a proper cutting list,' said the senior butcher. 'Assume that's OK?'

We didn't really have much choice. So we put the meat in the back of the car and headed home. The bags wouldn't fit in our fridge so we had to work on them straight away. They were due to go straight out to our customers but we were hoping for a rather more attractive presentation for our first produce.

Frances hunted in the pantry and found some Sainsbury's freezer bags and I carefully extracted the first cut-up lamb. It was OK but not brilliantly cut, but then I was in no position to judge at this early stage. We didn't have any labels so we bagged the meat up as best we could and scrawled approximate joint names on the outside in felt-tip pen. We then did the same with the other lamb and found we'd nothing to put the rebagged lamb in. I hastily found a couple of cardboard boxes and we packed one animal into each and set off to deliver them. Bill and Helen decided to leave us to it at this point, having witnessed the chaotic and disorganised start of our lamb retailing enterprise.

Luckily our first customers were friends who were happy to take our bizarrely presented produce on trust. Within days we had phone messages telling us how great our lamb was and reserving more for later in the year. We were clearly doing something right.

And so ended our first season. Ten ewes and fourteen lambs on our land at home. We'd learned a huge amount about sheep-farming and a bit about retailing but most important, we'd learned that we wanted to continue and try to build something viable and sustainable from our new hobby.

TEN EWES IS NEITHER sustainable nor viable and we'd enough grass for more sheep. I remembered from my sheep lectures that we should be able to get four ewes and their offspring to each acre and as we had eight acres, we ought to be able to run thirty ewes. Impeccable thinking but in the event quite difficult to achieve. I'd already bought another ten ewe lambs from Bert the previous autumn in case we decided to expand and on the same assumption that they could be sold easily if we decided to stop.

I made a mistake with these animals in that I assumed they'd

'The following day I took my haybob out and spread the grass so the underside was exposed to the sun.'

be OK outside being fed just hay, but they weren't. There's a difference between what you offer an animal and what they actually eat and the hay I'd bought in wasn't good enough. Lack of experience on my part meant I didn't realise how little they were eating and they got very thin. Good sheep farmers handle their animals regularly to assess their body condition but mine were in a distant field without a catching pen. Very poor excuses I know but luckily I realised there was something wrong and got them inside and onto some decent concentrate feed before any permanent harm was done. An important lesson learned.

The following summer, I decided it was time to make our own hay so we had more control over the quality.

We didn't, of course, have the equipment to make hay but I did now have a tractor, bought for me for my fiftieth birthday by Frances, and I scanned the local paper for what we needed. I found a hay turner – a device that goes on the tractor and can be used to spin out hay to dry it and then gather it up into rows so it can be baled. A delightfully named haybob which we collected from a chap in Wales one Sunday morning.

I decided that we couldn't afford a mower or baler this year and Frances concurred. So I persuaded a friend to lend me his mower and he agreed to come and bale the hay for me.

I scanned the weather forecasts and agreed a time to cut the grass when there was a five-day dry spell. I was working from home that day but sneaked out for a couple of hours to mow my first hay crop. It went amazingly well and I left the neat rows of silvery shiny grass shimmering in the heat of the sun.

The following day I took my haybob out and spread the grass so the underside was exposed to the sun. The smell was gorgeous, a sweet fruity waft born on a gentle breeze. I wished I could bottle it, and this feeling.

I carried on turning it each evening, mindful of the old country wisdom that you shouldn't turn hay too often or too violently or you'll break all the leaf off and be left with just stalks. By Saturday morning, I'd decided it was ready to bale and Jim arrived with his baler.

Immediately, the limitations of making hay with modern(ish) equipment on tiny two acre fields became apparent. He couldn't pick up the hay in the corners and since corners make up a high proportion of the area of tiny fields, we had a problem. I solved this, quite brilliantly I thought, by placing a family member in each field corner with a fork so they could pull the missed hay back in front of the machine. The family members, out in the dust and the full sun, didn't necessarily agree on my brilliance, but we got it done.

The family deserted me at that point, so I had to get the bales under cover by myself. I pitched them a few at a time and then clambered onto the trailer to stack them. It was then a trip of a hundred yards or so to the shed where I had to unload and restack them. Mercifully, the tiny acreage meant there was only a couple of loads and I returned to the house, covered in dust and sweat but insanely pleased with myself.

So at the end of 2006, we had twenty ewes to tup and Basil arrived again for his holidays. We followed the same routine and at scanning we were delighted to find that all of them were pregnant. We also achieved the remarkable feat of delivering, rearing and selling all the lambs we scanned. Losses of zero percent are unheard of in the sheep world and we never achieved it again but it was a nice feeling and gave us confidence that we could add another ten animals to our flock. This time I persuaded Bert to sell me yearling ewes. In the surreal world of sheep-speak, 'yearling ewes' are actually two years old and ready to have their first lambs. They're much more expensive than ewe lambs but they can hit the ground running, as it were, and become productive straight away.

This meant we were now at our target of thirty ewes for lambing in 2008. We were developing a nice little hobby business. It was great fun, we got to share it with our friends Ian and Jill at lambing time and the family seemed to enjoy helping out – except at haymaking. I had my tractor and the other toys to enable me to make hay and we'd a growing band of customers ordering our lamb. We even had a waiting list of potential customers which was great for our confidence.

But then something happened that turned our world upside down. A huge change in lifestyle and the opportunity to turn this 'nice little hobby business' into something more viable and possibly even profitable.

IT WAS A WET FRIDAY afternoon in Swindon. Days like this were miserable anywhere but Swindon somehow added an extra edge to the melancholy. I was sitting in a hotel foyer staring blank-eyed at a cup of coffee, trying to decide whether I deserved a slice of cake to go with it. My ADAS marketing mentor Paul was sitting opposite me, smiling. He was always smiling, but then if I was being paid what he was to work with managers like me, I'd be smiling too. No responsibility, easy-going chats, a few critical remarks and an action plan and a nice big fee.

We'd spent the afternoon going over my sales plans and cash flows and they were OK. Nothing spectacular but OK. We'd meet the targets, we might even exceed them, 'but not by too much,' he warned, 'not after last year!'

In the inexplicable world of venture capital-backed business, exceeding your targets is as bad as not reaching them. Apparently my budget-busting figures the previous year suggested a lack of control.

Paul looked across at me, his smile fading slightly. 'This really doesn't float your boat does it Andy?' he asked cautiously.

'No, but it pays the mortgage,' I replied. 'It's not what I joined up to do all those years ago. I'm an agriculturalist, not an environmental scientist, but I'll have to stay put for a while.'

It was true, ADAS had changed out of all recognition in thirty years from a farm advisory service to a consultancy company delivering 'environmental solutions'. Many of the scientists in my team knew nothing about farming and indeed some even saw farmers as the enemy of their world-saving, crusading green agenda. I wasn't a dinosaur but I was uncomfortable with this wholesale dash away from the industry that had been my life and that I loved.

I realised Paul was speaking again. 'You know as well as I do that there's a tsunami of change coming towards you. Your group will get merged with two others and there'll be three bosses for one team.'

'You don't have to tell me that,' I said crossly. 'I was on the group that recommended the merger!'

'So what are you going to do then? Fight for the top job, which you don't really want, 'cos its 'Environmental', or do what most ADAS people do and let the tsunami wash over you and hope there's a job on the other side? You know you could just take control, walk away and do something else.'

'With a daughter at university and bills to pay, I can't afford to,' I replied but a bit less convincingly.

'I'm serious,' said Paul. 'Why sit in a job you don't like for the next twenty years, doing the same old thing month after month. Keeping the board happy with an annual review and forever chasing the next contract from customers you don't even like very much?'

'Put like that, it's a pretty underwhelming future I've got, isn't it?'

He laughed, the smile was back. 'Well, at least think about it, OK?'

I made my way home through the busy Friday traffic, the rain had stopped and the roads were glistening in the late afternoon sunshine. I dared to allow myself to think what it might be like to move on. I could do some more farming maybe? I could do some private consulting work? How much money would we really need? The thought of leaving was very attractive. I decided on that journey home to talk it over with Frances over the weekend.

After hours of discussion, we concluded that maybe I could agree to leave, assuming they'd let me (which they would – I wasn't that special and I cost a lot to keep) but ask to stay on until the following summer when I would stop paying university fees. So I spoke to Colin the MD on Monday morning. 'OK,' he said, a little too quickly, 'we'll see what we can do.'

Two weeks later my boss, Martin, called me into his office. He was supposed to be on leave but he'd come in especially. He was a close friend but he was talking quite officially.

'Hang on a minute,' I said in alarm. 'Are you serving me redundancy?'

'Well yes,' he replied, 'as you agreed with the MD.'

'But I didn't agree anything,' I gasped, 'just that I'd be willing to fall on my sword as long as it was mutually beneficial – and definitely not this quickly.'

Silence. My colleague looked uneasy. 'Well I'm supposed to be on leave – you'd better sort it with Colin.'

I guess having redundancy served on you is not a great feeling at any time but it's especially unnerving during the summer holidays when there's nobody to talk to, nobody to discuss details with. I eventually got to talk to Colin a week later.

'I agreed to go if it was mutually beneficial,' was my carefully rehearsed opening line.

'OK, so what's the problem? Is it the money?' He'd clearly been briefed by Martin, and possibly Paul.

We discussed and agreed a package and I was to leave at lambing time the following year, with paid leave until May. It was all very amicable but their urge to get rid of me was unnerving. Was I really that expendable?

My last day in ADAS was rather pleasing – I'd been asked to speak at a conference in London about the development of Farm Computing in the UK – something that I was uniquely qualified to talk about. When I arrived, I found that many of my old friends from the industry had come along as they'd heard it was to be my swan song. I had a fabulous day, reminiscing with them over lunch, and made my last presentation as an ADAS man. As I drove down the M4 on my way home, my watch beeped at five o'clock to signify the end of my ADAS career. I glanced out of the window to see I was driving past the Bracknell turn.

Granted, Swindon and Bracknell are not especially promising starting points for a full-time career in farming, but in this case, that was exactly how it happened.

3. FARMING BY ACCIDENT

'*He who tells you to follow your passion was probably already rich.*'
– Prof Scott Galloway

IT SEEMED LIKE A GOOD IDEA at the time. I could ease
gently from my ADAS job into a more relaxed mix of private
consulting, a bit of teaching and have more time for farming.
Nothing huge but we could maybe take on some rented grass
and have a few more sheep and perhaps we could even bring
the butchery in house and save some money. But of course it
didn't work out like that.

The teaching never happened. I applied for several part-
time posts at agricultural colleges but never got a response, let
alone an interview or even a discussion. I thought some broad
business experience, a lifetime in ADAS and an agricultural degree would have been useful. I'd done some visiting lecturing

over the years and always enjoyed it. The students had been complimentary and the feedback had always been good but this clearly wasn't going anywhere.

The consulting did at least get off the ground, just. I'd booked in several jobs before leaving ADAS and was ready to embark on the first when I got the call, 'Sorry Andy, we're having to cut our budgets, this talk of recession is getting serious, we'll be in touch.' I did some small contracts including a project for a former colleague who was building a cost comparison website for farmers but the rest of the work dried up as the recession hit. Within a couple of months I'd exhausted all the options and ended up with just the farming to keep me occupied.

And so that's how it started. Before I knew it, I was farming full-time – literally farming by accident. And the first job was to figure out how to move this hobby business up a level into something that at least had a chance of making some money.

THE CHAIRMAN OF ONE of our local networking groups had been badgering me for a while to come and talk to them about turning my hobby into a business. I'd joined this group hoping to develop my consulting work but they'd turned out to be better customers for our lamb. Now, as a full-time farmer, I decided a talk would be great publicity, but as I sat down to prepare it, it occurred to me that there are fundamental differences between hobbies and businesses. Differences that we only became aware of with hindsight as they became a hindrance to our progress.

I reckon the biggest difference between our hobby and our business was the focus. As a hobby the primary aim was to have fun and it was about using our time enjoyably. As a business we needed to make efficient use of our time and make money.

In my case it was the farm machinery. I toured farm sales

'I toured farm sales buying ageing equipment cheaply.'

buying ageing equipment cheaply – a mower and baler that were older than me, a bale loader which was only really worth its scrap value and endless other bits of dilapidated kit. 'Deferred rust' as an ADAS colleague used to call it.

I liked tinkering with machinery. My Dad was an engineer and I think I got it from him. Unfortunately, as the business grew, this strategy landed me with two problems;

Firstly, this old equipment just wasn't good enough to use on a commercial scale – the mower and baler were too slow and as we shall see later, the tractor wasn't strong enough for the tasks I asked of it.

The second problem, keeping it working, was at best time-consuming and on some occasions, downright damaging to the business.

I was battling against the weather. I'd a field full of straw bales to move and it was going to rain that evening. I'd got myself organised with two trailers of my own and a borrowed one which would take most of the bales and I had a sheet to cover what was left in a stack in the field. My elderly tractor was more than capable of loading straw bales, with a grab on the front, eight at a time onto the trailers and I was sure we'd get it all done before the rain arrived. But halfway through loading the first trailer there was an almighty bang and the tractor front wheels ended up pointing towards each other – in the 'snowplough' skiing position. The steering arm joint had broken. I made a hasty call to the local parts department and no, they hadn't got one but could get one for the next day. But it was going to rain. The bales were scattered around the field, impossible to collect and sheet by hand, so I called Frances to fetch me home and left my entire winter stock of bedding straw to get soaked by the rain.

If we'd started out to run a financially sustainable business, we'd have spent our limited money on productive assets rather than machinery or 'the toys' as Frances insisted on calling them. Ewes produce lambs which have a value and can be sold; mowers and tractors don't. We could have borrowed or hired machinery for the field work and used contractors for hay and silage – or even bought in the forage ready-made.

The other difference was one of scale – or more accurately, aspiration of scale. In our case, we never intended to operate as a full-time business and our aspiration in those early days was a small operation supplying lamb to our friends and family. This gave us problems when we tried to expand.

I was at one of our local farm sales. Spring had finally arrived as I stood with the sun on my back, watching the punters nosing round the various lots. I was getting quite well-known now and on nodding acquaintance with many of the bidders. They rarely spoke to me but I did hear one of them confide in his mate, 'It's that bloke again – you know, the one who buys the old stuff.'

I was bidding on a small stock trailer and was delighted when I got it for just £200. I couldn't understand why there were so few people bidding but as we expanded the flock it became all too clear. This bargain trailer was only capable of carrying eight ewes or twelve lambs, which was fine for the hobby flock based at home, where all we needed to do was move a few lambs to the abattoir or fetch a borrowed ram once a year.

But as we grew the business and started using rented ground, the logistics became more challenging with endless journeys back and forth using scarce time and expensive fuel. One notable Christmas eve we spent an entire day moving just thirty ewes home, eight at a time, in the snow.

We should have started big. Growing from small beginnings seemed like a safer bet but it wasn't the right thing to do. If we started again, I would buy as many ewes as I could afford, a decent trailer and vehicle and borrow or hire everything else until we were established. As it was, we had a much slower and more frustrating path to a sustainable and profitable business.

IF WE WERE GOING to grow, we needed more sheep. We'd been buying them from Bert ten at a time and our flock had grown to thirty but this was far from a commercial operation. So I decided to go to market.

'If we're going to do this properly, we might as well lamb three hundred as thirty,' was the impeccable wisdom I tried to impart to Frances. She remained sceptical. 'Where would you put that many sheep? We haven't got the buildings or grass to keep them?'

'The little trailer will only take a few at a time so I can't buy more than a dozen or so and that would mean two trips. I just want to go to market and see what's about.' I sounded more confident than I felt.

And so the following week I drove into the car park at Hereford market and carefully parked my ageing car and even older stock trailer next to the gleaming trucks and huge aluminium stock boxes that most 'proper farmers' seemed to have. It was a glorious morning, bright October sun and pleasantly warm as I walked across to the sheep pens.

Rows and rows of hurdle enclosed pens, each containing sheep peering expectantly through the bars. There were over three thousand lambs penned up for sale that day. Making their final trip to the abattoir or, for the lucky ones, maybe a trip to another farm until they grew big enough. But I was looking for the ewes, hopefully those being sold because the farmer was retiring or maybe changing his business. I didn't want 'cull' ewes that were past their best and I couldn't afford the young ones that 'proper farmers' would be buying as replacements for their own flocks.

I cautiously made my way to the office to pick up a catalogue. The girl behind the desk eyed me suspiciously. 'You've not bought here before have you?'

'No,' I said. 'Is that a problem?'

'Only that you'll need a buyer's number as we don't know you and we need your bank details so we can make a phone call to make sure you're who you say you are.'

'Well so much for friendly local auction marts,' I thought as I handed over my details. The girl pushed a tattered piece of card at me with a number on.

'Just wave that at the auctioneer if you buy anything and we'll know who's bought that lot,' she muttered.

I now had to walk back to the car to find a pen. I really wasn't in my element here. I'd been to lots of livestock markets but I'd never actually bought anything before.

Trying to look as if I knew what I was doing, I walked up and down the lines of sheep pens making notes in my catalogue. The animals were beginning to pant in what was becoming a warm sun but they'd got a long time to wait yet. There were Suffolk cross ewes with jet black faces and lots of so-called mules that were crossbred sheep from Wales or northern England. I was looking for Suffolk cross animals, not because I had any special preference but because that was what Bert had already sold me and I figured it would be simpler to stick to one breed or cross.

I found two pens of three and four-year-old sheep, assuming the catalogue was correct – I wasn't clever enough yet to assess sheep's ages from their teeth. They looked OK to me although I really had no idea what I was doing. I marked these two lots in the catalogue and went for a coffee and bacon roll, which seemed to be what everyone else was doing.

An hour or so later, the auctioneer was making his way precariously along the boards across the backs of the pens. Laughing and joking with the buyers and sellers as he did so, taking no nonsense but keeping up a steady stream of banter between lots. Everybody seemed to know everyone else but all I got was the odd curious look. The auctioneer reached the lot before the one I wanted. I moved towards the pen but was pushed aside by others who clearly thought they had a greater claim to be there than me. But I persisted and positioned myself where the auctioneer's assistant could see me.

'Right then we have a pen of thirty-five nice three-year-old Suffolk crosses here – Severnoaks.'

A man climbed up onto the boards and exchanged a few bawdy pleasantries with the crowd. I assumed this was Mr Severnoaks but my mind was focused on the fact that there were thirty-five of them. How will I get them home? It'll take all afternoon. And more pertinently, why hadn't I counted them?

The bidding started lower than I expected but there was one keen buyer, a genial older man who smiled at me as he bid against me. We kept going and then stopped with his bid at £74. I couldn't afford to go higher.

'Go on,' shouted Mr Severnoaks, directly at me, 'you know you want them!' The crowd laughed.

I nodded at the auctioneer, he took my bid and the genial man smiled at me.

'OK you 'ave 'em,' he laughed.

I was shaking with the tension, I'd just bought my first thirty-five sheep at auction and I didn't dare try and work out what thirty-five times £77 was, but I knew it was a lot.

We then moved to the four year olds.

'Twenty-nine of these,' called the auctioneer.

The bidding started at $60 but was very slow. It stumbled uneasily up to seventy but Mr Severnoaks refused to sell.

'I'll take 'em home again at that price,' he growled, the smile fading from his face.

Then I had an idea. As he clambered down from the boards and came over to give me some 'luck money' – a tradition that has all but died out now except it seems in Herefordshire – I ventured rather timidly, 'If I pay you £75 for the four year olds, will you deliver them and the others to my place for me?' He thought for a moment.

'Course we will,' he laughed. 'How far away are you?'

We did the deal and shook hands.

'Go and pay for the three year olds and then come back here and write me a cheque direct for the rest,' he commanded.

As I walked back to the office to pay, I got a few smiles from fellow bidders. The genial smiling man came up and clapped me on the shoulder, 'You obviously wanted them, din't you,' he said. 'They're nice sheep and Severnoaks is fair, they'll do you well.'

I was beginning to feel a little less of an imposter as I queued up to pay but as I reached the desk, the same girl scowled at me as I took out my cheque book. She sat there with her hand out-stretched.

'We write the cheques so we can read them,' she muttered – another tradition I was unaware of.

When I got back to the pens, Mr Severnoaks was supervising the loading of my sheep into his lorry.

'Come over here,' he called. 'I don't want this lot seeing my business.'

I wrote him the cheque hidden behind his truck and we shook hands again.

'Can't make you out,' he smiled, 'You're not a farmer are you? You look more like an accountant or solicitor to me. Still, your money's as good as anyone else's, probably safer than most of this lot.'

We set off in the lorry and after forty minutes or so, we arrived in our yard. His driver expertly backed in and unloaded the sixty-four sheep into some hastily erected pens. With a laugh and a wave he drove out of the yard in a cloud of October dust and I turned to view my new charges.

What on earth had I done? Our flock had gone from a hobby-sized thirty sheep to almost a hundred in one step. Still not really a commercial scale but a good way towards it. The animals were glad to be away from the heat of the market pens and many were taking long draughts of water and nibbling interestedly on the hay I'd found for them.

At that moment Frances arrived home and drove up into the yard. As she made her way towards me, I started preparing my defence. I also needed her to drive me back to Hereford to collect my car.

'They're a good deal, I got them delivered for free, lots of people were after them so they must be OK!'

Frances got out of her car and gasped at the sight of the pens full of sheep. I grinned sheepishly and to her eternal credit, she smiled...

AS FRANCES HAD FORESEEN, more sheep needed more grass. New, would-be tenants in an area have to prove themselves – not just that they have the cash to pay but that they are reliable and will look after the land and their animals properly. Decent rented grazing is always exchanged by word of mouth and as newcomers we weren't in that network. So, we had to start small and take on pieces of ground that other, larger operations didn't want.

The first proper rental we took on was at Leysters, a nearby village. I was sitting in a chair one Sunday morning in Edwin's 'front room'. He was an amiable man, well into his seventies, known to everyone as the former proprietor of the local hardware store. His brother had farmed the land but had recently died and his cousin ran the pub across the road. 'It's mowing grass,' he said. 'It'll be £60. Just take a cut of hay off the two back fields and I'll keep the cattle in the front one.' Beneath the geniality was a sense that it would be pointless to try to haggle, so I agreed. It was only ten acres in total and the fields were tiny but it was a start.

The arrangement worked well and we got some good hay off it, so I decided we'd have it again the next year, by which time Edwin had sold the cattle. We secured all of Edwin's ground by offering a little above the going rate on a ten-month grazing let and grazed ewes and lambs up there from March to December every year from then onwards.

I liked Edwin and we developed a good relationship with him. He was a decent man, if a little eccentric. He was also a shrewd businessman and sometimes his cautious nature where money was concerned got the better of him.

One year we did a deal to improve the fencing on his fields. I'd had one too many excursions to retrieve sheep from a neighbour's land and we agreed that I would buy the materials and he would pay the contractor. All went well until the end of the job when I arrived one morning to find Tom, a very disgruntled contractor, trying to straighten old staples to finish

'As Frances had foreseen, more sheep needed more grass.'

the last few metres of fence. 'I told Edwin to let me know if you needed more,' I said.

Tom scowled. 'I did but he went down the field last night and counted how many staples we used on each post and how many posts we'd got left. Then he emptied the bag and counted the staples and decided we'd got enough with one to spare! Unfortunately we brought a couple of extra posts this morning to tidy that gap up.'

I started to laugh. 'Sorry Tom but that's Edwin all over. He can be frustrating but he's a very good landlord to us.' Sadly, Edwin is no longer with us but I can still see him smiling through his kitchen window, cigarette in hand, watching me on a tractor and waving as I pass his garden gate.

We added other blocks of land as they became available. One was a small six-acre paddock with a really useful barn where we wintered the fattening lambs. We spent many Monday mornings in that barn, weighing and selecting lambs to go to the abattoir. The problem was it was too far away – nearly thirty minutes on a tractor – so it didn't help our quest to become more efficient. As we got bigger, we took on larger blocks

'...much of the land was on a very steep slope.'

of land closer to home and I would have given up that single field if it hadn't been for the barn.

The other small rental was a thirteen-acre block owned by a friend. It was a nice, large (for us), level six-acre field where we could make hay and silage. There were also two grazing paddocks including an orchard with a small field shelter known for some reason lost in the mists of time as the 'Rondavel'. This was a perfect place to turn out ewes and lambs early in the season. The grass grew early under the shelter of the apple trees and the Rondavel was the ideal refuge for small lambs when the weather turned wet.

Our first proper commercial-sized rental agreement came as a result of a chance meeting in the pub. Adam was primarily an arable farmer but he had thirty acres of grass available so we took it on and doubled our acreage overnight. We built a catching pen to enable us to handle ewes and lambs more easily and at last we could have large mobs of sheep grazing together, which reduced the time we spent travelling between sites. The downside was that much of the land was on a very steep slope which made gathering them very hard work. We were very grateful to Adam for this chance to 'step up' to bigger land blocks and we made this arrangement work for a few years but were always on the lookout for an easier-to-manage block closer to home.

Over the years we painstakingly built ourselves a good reputation by paying rents on time and looking after the land we took on. And as we became accepted, negotiations became easier.

THE FACT THAT OUR rented grazing was mostly on annual summer lets meant the sheep had to come home in the winter. With only eight acres of grass of our own and more than a hundred sheep by now, we needed to be able to house them to avoid turning our tiny acreage into a mud bath. I have to admit the barn was an extravagance really but I justified it, to myself at any rate, as adding value to the property.

Still fixated with trying to save money by doing things myself, I had hired a digger and dumper to level the ground and members of the family were drafted in to help.

This was the time when Kevin entered our lives. He worked for a local ground works company and was employed by the barn supplier to dig the footings for the steel frame. He arrived with a smile, in his shiny yellow JCB, with a real can-do approach and his skill was amazing – he probably could peel the proverbial egg with that digger and he did a great job, very quickly. We hit it off immediately and he became a great friend and helped me with many projects over the years.

With the footings done, it was time for the erectors to do their bit. Joe and his dad, Stan, bounced cheerily into our yard one Monday morning and Stan's first words, before he'd even got out of his van were, 'You do know it's too small, don't you?' This was an eighty foot by thirty-six foot barn replacing an old ramshackle shed put up just after the war. 'It's huge,' I

'Getting the barn up and the walls done was a huge achievement for us.'

said. 'We'll never fill this.' But of course it was too small and we did fill it, very quickly.

Stan and his lad put the frame up and had the roof on within a week and finished the side sheeting and doors a couple of days later. We then had to do the rest. We built eight-foot-high reinforced concrete block walls which took months of work. I bought scaffolding and planks to work on, and we used the tractor to lift pallets of concrete blocks up to where we needed them. We then had to fill the blocks with concrete as we went and the whole job became very hard work indeed.

Frances spent hours standing on those scaffold boards with freezing toes and fingers, pointing the blocks and brushing them down and helping me pour concrete into the finished walls. Help that I'm eternally grateful for but that was way beyond reasonable for me to expect.

Getting the barn up and the walls done was a huge achievement for us but we still needed to install a water supply and electricity. Here the challenges were more bureaucratic than physical as we battled with those who sought to deter us.

The battle was with officialdom. It was agreed after some debate that we'd have a dedicated water connection but we would run an electricity supply under the lane from the house. The difficult bit was persuading Welsh Water to come and make their connection while we had the road dug up to lay the cables. A nice lad called Gareth agreed it would be a good idea – 'All nice and joined up,' as he put it – but then decided, in late January, that coordinating the two operations on the same day was beyond him.

'Are you seriously telling me we'll have to have two trenches dug across our tiny lane on two different days?' I fumed. 'Do you really, honestly, think that's sensible?'

'OK Andy, I'll see what I can do,' he muttered, 'but I doubt they'll agree.'

But they did, eventually, and amazingly it worked.

OUR BUSINESS WAS GETTING bigger. We now had more sheep and more grass to graze them on and a much bigger building to keep them (and us) warm and dry in the winter. But the one thing we didn't have more of was time. Time to check the sheep, to minister to their every need and of course, to move them around. I wasn't aware how often sheep needed to be moved until I started farming.

They say you should never move sheep with your wife and we proved that to be true very early in our farming career. Frances was (and still is) a hugely intelligent woman who I loved dearly but she lost every bit of the common sense she ever had when we were moving animals around. And she said the same about me. So to make moving the sheep easier (and to preserve our marriage), we decided to buy a sheep dog, probably one of the most significant pieces of 'equipment' we ever bought.

The following story is from much later in our farming career but it sums up perfectly the contribution the dog made to our business and indeed to our lives. The rain was coming down in cascades, the sheep were soaked, as was I. The gate on the trailer was stuck and I couldn't get the animals off the upper deck and as the rain dripped down my neck, I realised the dog had gone. Unbelievable, I thought, just when I needed him most and he's gone off into the dry somewhere. The sheep from the lower deck pushed past me and disappeared down the loading ramp, 'Geri!' I shouted but he'd gone.

Geri (short for the Welsh name Geraint) was my dog. He'd arrived with us at two years old, supposedly a fully trained sheep dog but with very little experience. That coupled with me as a novice dog handler meant that we didn't always see eye to eye. Indeed we'd lost him completely on his second day with us. He decided he didn't want to work with me, slipped under the gate and was gone. We scoured the countryside for three days, putting up signs, asking locals and running a huge email and social media campaign but to no avail. And then one morning there he was, back in his pen, looking very sad but resigned to staying with us.

He lived outside in a pen as that was what we'd been told to do. 'Don't make them soft,' was the advice, 'and remember, they'll only work for one master.' All well-meaning but completely wrong. Geri was a border collie, supremely intelligent and perfectly capable of understanding when he was working and when he was being a pet. He also understood commands from both of us although he had a highly developed sense of priorities and would only do what Frances asked if I wasn't around.

It soon became clear that he had a much better grasp of what was required than I did and I very quickly realised that I should leave him to it when we were moving sheep. He seemed to understand what the sheep were going to do before they did and could bring them to me from a great distance away with only the simplest of commands.

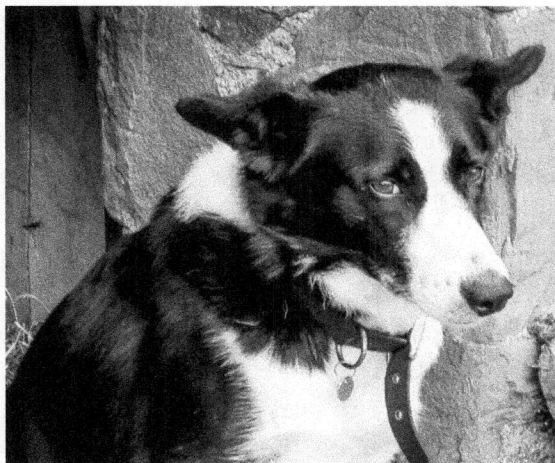

'Geri (short for the Welsh name Geraint) was my dog.'

His clairvoyance wasn't always an asset however. On several occasions I'd be in the process of putting up pens and hurdles when he'd appear with a batch of sheep that he'd gone off, unbidden, to fetch. The resulting chaos as they scattered the hurdles, before disappearing back whence they came, took precious time to fix.

It was hard to get cross with him though; he was so keen to please and saved me hours of walking and chasing, particularly as the flock, and the fields, got bigger. And like all collies, he was motivated solely by praise. We never once gave him rewards but over the years he became part of the team and eventually moved into the kitchen.

So on this occasion I was surprised he'd let me down. 'I must stop letting him into the house,' I muttered. 'He's getting cocky and starting to take liberties. I'm soaked to the skin and he's no doubt curled up by the fire. He's probably even persuaded Frances to give him his tea!'

I finally managed to get the gate open to release the sheep from the top part of the trailer and as they scampered off up the field I wearily trudged across the yard to look for the rest. And there he was, even wetter than me, crouched by the gate to stop them going out onto the road. He knew that if he'd tried to bring them back they'd have scattered and he'd have lost some, possibly into the path of oncoming traffic so he'd stayed put until I arrived. It was probably instinct but with a level of intelligence and foresight that we had come to value very highly.

It's no exaggeration to say that we couldn't have built the business without Geri. His tireless loyalty and good nature were always a pleasure to work alongside. He was also great company on long days out in the fields and equally long nights in the lambing shed.

Sadly he's no longer with us. He managed sixteen years which is a great age for a collie and he spent his final months asleep in front of the fire or standing and staring blank-eyed into space having forgotten why he'd got up in the first place. Characteristics his master is taking on all too quickly.

THE FARMING WAS PRETTY straightforward, my knowledge was a bit rusty and I needed to update my practical skills but I broadly knew what I was doing. Retailing, on the other hand, and hopefully making some money was a different game altogether. 'Sheep-keeping is a numbers game,' I mused to my friend Ian during a late night in the lambing shed. 'The profit on each lamb sold is a few pounds at most so you need lots of them to make a living.'

'Well get some more then,' was his not unreasonable response. 'You've got the barn space and presumably you can rent some more grass?'

'We'd need to scale up dramatically though – three to four hundred ewes and triple our acreage. At £100 plus apiece and £80 an acre, we'd be looking at over forty grand for the sheep and another twelve for the rent annually – even if we could find the grass.'

'So what are you going to do? Treat it as a hobby and lose money? That doesn't sound like you!'

We'd already shown that we could sell lamb directly to our friends and I'd been looking at the figures. I reckoned we could scale that part of the business up.

'If I take these lambs that are being born now to market when they're fat in October, they'll fetch about eighty quid,' I continued. 'At the moment, if I have that lamb slaughtered and cut up and sold as meat, it'll be worth twice that.'

'So you're saying that about half the value of that lamb is taken by someone other than you?'

'Yep, at least half – more value is created 'outside the farm gate' as we say. And that's why I think we should focus on selling our lamb direct to our customers, so we get all the value for ourselves.'

'Makes sense to me, but it'll be a lot of work. How does Frances feel about it?'

I don't think Frances realised what we were letting ourselves in for and the truth is, there was another, more personal reason for setting up the retailing business. I wanted us to be taken seriously by our farming neighbours and I was sensitive to the view that we were 'playing' at farming and we didn't really need to make it pay. So, setting up a proper business, producing and retailing our own lamb would, I hoped, establish us within the farming community as doing something worthwhile and a bit different.

Our first venture into lamb retailing, in those early days had been a fairly chaotic attempt to provide friends and family with whole or half lamb packs. This was the easy way to start as we didn't get left with the less popular bits that people didn't want. It's what butchers call 'balancing' the carcase – which is a big part of making meat retailing profitable. But as we expanded, we found that our new customers often couldn't afford or hadn't the room for large quantities of meat, but they did want a half leg of lamb or a pack of chops.

Initially, we met the demand for individual cuts by 'borrowing' them from the whole lamb packs, but this was doomed to fail. It took me several months to realise that we could store the cuts separately and sell them individually or make up whole and half lamb packs as we needed them.

As part of the drive to reduce our costs, at this point we decided to bring the butchery in-house. This undoubtedly saved us money on processing charges but it involved investment in a butchery and cold store. It also involved me attending a two-day butchery course in Sussex, which was great fun. We decided to ask Morag, our Environmental Health Officer (EHO), to help us design the butchery, figuring that she'd be more likely to accept premises that she'd helped to plan.

She helped with the layout, advised on the right paint to use and explained why we needed the two sinks (one for hand washing and one for utensils). It was a fully insulated room with its own plumbing and water heater and its own separate electricity supply. We laid a special concrete screed floor so it could be scrubbed and used special scrubbable paint on the walls – all to Morag's specification. We installed a proper butcher's block that I bought on eBay and put in stainless steel tables to pack on and shelving so that everything else (packaging, labels, meat ties etc.) could be kept away from the clean surfaces.

In fact it was pretty near perfect. The only trouble was, it was too small.

Yet again the hobby mentality was slowing us down as we got serious. It was fine for processing our thirty lambs for friends and family, but when we had three hundred and fifty coming through every year, it was barely adequate.

'At this point we decided to bring the butchery in house.'

We also needed a cold room – a bit like a big 'fridge that you can walk into. We needed to be able to hang lamb carcases prior to butchery and have somewhere to keep the butchered meat before it went out to customers.

Generally, our routine was to take the lambs into the abattoir on a Monday and fetch them home on a Friday. We then hung them in our cold room until the following Wednesday or Friday, so in total they were usually maturing for at least ten days.

My butcher friends weren't happy. 'You can't do that!' they'd cry. 'They'll go off or go mouldy.' But our customers proved them wrong.

Sally, a valued and lovely friend and a brilliant cook had bought a whole lamb from us which I left in the cold room over a weekend, intending to cut it up on the Monday. Unfortunately I forgot all about it until the Friday by which time it had been fourteen days on the hook. It appeared OK and I hadn't time to get another one so I decided I'd better cut it up for her. A couple of weeks later I got the call. 'How long do you hang your lambs for, Andy?' she queried.

'Oh... about ten days, why?' I tried to sound nonchalant.

'Well, that last one you did for us was the best lamb I've ever eaten.'

Our USP for our produce became our guaranteed hanging time. Most butchers only hang sheep meat for a day or two and supermarkets seldom hang theirs at all.

BACK TO THE COLD ROOM. After much searching and several false starts, I found someone who worked for a large company installing refrigerated potato stores. They used special interlocking panels to build these stores and often had 'a few left over'. He had a very nice little weekend business supplying these oddments to people who were doing exactly what we were doing and for a couple of hundred quid, he'd even come and install them for me.

'I suppose you'll want a door,' was his parting shot. 'I'll see what I can find.'

It was at this point that I made one of the biggest mistakes of our entire Whyle House Lamb journey. I noticed an advert in the *Farmers Weekly* for a 'fridge box' – literally off the back of a lorry – a few miles from our home. In its former life, it was a Ginsters Pie truck. This was an easy option. We could set it up outside the garage, it was ready-made, with hanging rails and a refrigeration plant, and if I could get it for a sensible price it would be a great solution. The seller was someone I knew slightly and he immediately dropped the price by nearly fifty per cent when I called, which should have made me cautious.

'The fridge box remained in our yard for some months until I sold it again.'

'Us local guys have got to support each other,' was his reasoning.

Since it had been removed from a truck, the pipework had been disconnected but he assured me it was easy to fix. He even put me in touch with someone who gave me a quote to get it running. All in all, it looked like a good deal and when he offered to deliver it free of charge, I agreed to have it.

It arrived in our yard and the first problem was it was too heavy for my tractor to lift off the lorry. We managed to move it eventually and dragged it to the side of the yard. We quickly realised that it was going to be very difficult to move it down to the garage where the butchery was but decided we could still use it in the yard. Dave the fridge engineer duly arrived in his van. He was a cheery chap who clearly knew what he was doing but he kept asking, 'How much have you paid for this?' implying that we'd not had a great deal. He got it going but the costs were astronomical.

The final straw was it was very noisy, far too noisy to have nearer the house, even if we could have moved it, and the Envi-

ronmental Health folk were very unhappy about us carrying carcases and fresh meat backwards and forwards across the lane!

So I went back to my original plan – a few spare panels and
£200 to install it, plus a second-hand fridge unit which I found
on eBay and we ended up with a really good, commercial style
cold room for £1,500 – about what I'd paid for the non-work-
ing fridge box before the repair costs.

The fridge box remained in our yard for some months until
I sold it again, a monument to my headstrong and foolhardy
purchase. Some hard lessons learned.

ONE OF OUR EARLY helpers in the butchery was Tomos. He
was the fearsomely bright son of a neighbour who for some
reason took to butchery and loved the intricacy of it. I taught
him to do simple jobs such as boning lamb breasts and cutting
up meat for dicing and mince, but he very quickly progressed
to the more serious stuff. He was unusually tall and bespecta-
cled and I can still see him crouched over the block (which was
too low for him really) carefully preparing chops and steaks as
if his life depended on it.

Tomos was also great at putting together our 'Quarter
Lamb' packs. This was another brilliant initiative of mine that
seemed like a good idea at the time. Giving the customers a
great deal of £40 worth of lamb for a fixed price of £35.

Unfortunately, the butchery was difficult as, rather unhelp-
fully, lamb carcases tend to break down into ten loin chops
and fourteen cutlets – neither of which divide by four. In other
words, we couldn't get four quarter lambs out of one carcase
without some imaginative and time-consuming cutting. Tomos
used to love gathering the components together to meet the
minimum value promise while not losing money and he even
asked me one day to cut the chops a bit smaller! He never got
it wrong in the three years he packed for us, but once he'd gone

off to university we decided that the time involved was really not worth the extra sales, so Whyle House Lamb Quarter Lamb packs were no more.

Our next idea was to try for some wholesale outlets, so we decided to go for lunch at a prestigious local restaurant, to see if we could persuade them to buy our lamb. We had a great meal and sent our card through to the chef. To our delight he came straight out to see us full of enthusiasm. But he wanted to buy whole lambs. 'I can do my own butchery, I just want to buy good local lambs at wholesale price,' was his position. We demurred. There was nothing in it for us, we always ran out of lamb in the summer and I couldn't see the point of selling lamb at a lower, wholesale margin when it would earn us more as a retail sale.

More successful was the deal with our local pub where they took our lamb for mid-week meals and for their Sunday carvery. We provided them with promotional leaflets and we did an annual joint promotion where pub customers came down to the farm to buy lamb at a discount. This was very successful while it lasted, apart from one glitch where we found the pub owners selling cheap cash-and-carry lamb as ours and had to have a few words about provenance. We also took on a big regular order to supply two and a half lambs a week to another pub and although time-consuming in the butchery, it did provide very valuable regular cash flow.

As a general rule the more prestigious an establishment, the less they will pay, so we avoided the real top-end restaurants with one exception. I knew the chef, he gave us great publicity at shows when he was demonstrating, but in all honesty, selling to him was better for the ego than for the bank balance.

MORAG WAS NOW HAPPY, we had a butchery and cold store that passed muster and we were at last in a position to start selling at our local Farmers' Market.

'It was with some trepidation that we approached our first Farmers' Market.'

'You can always tell craft fair people from Farmers' Market traders,' muttered David, the Market Manager. 'Craft people are always sitting down reading the paper whereas Farmers' Market folk will be on their feet, talking to their customers and selling.'

It was with some trepidation that we approached our first Farmers' Market in Leominster on 14th November 2009. We'd been interviewed by David, who had pronounced us suitable to trade. He visited us at home to make sure we were producing what we were selling and not buying it in and he told us in no uncertain terms that there were rules we had to abide by. No arriving late, no leaving early, help set up the stalls and while trading absolutely no (on pain of death) sitting down!

On that first day we met Enid, a ferocious trader selling vegetables and fruit, loved by most of her customers, even though they usually ended up with several cauliflowers rather than the one they needed, as one of her 'deals'. She had that sparkly, almost impudent look which said, 'I'm going to sell this to you, no matter how much you resist.' I learned a lot from Enid about engaging with the customers and 'up-selling' to increase the sale value for each customer. One of the best lessons she taught me was when customers ask to see a pack of produce (such as chops, in our case) always present them with two or three options. Often they can't decide which to have and will take them all!

David introduced us to the other traders when we arrived, including many who became lifelong friends and they made us very welcome. It was freezing cold but we had a great time and took £102. This was a turning point for us as we realised we now had a way of selling our produce to the public.

A bit like farming by accident, we'd stumbled on a way of moving our retailing up a gear.

The move to market trading meant a major change in our processing routine. Firstly, we had to decide what to take to each market. Should we take half legs or whole ones? Should we bone and roll the shoulders? Loin chops were straight forward but cutlets could be cut lots of different ways as well as being boned out and rolled to make a lamb fillet.

Another valuable lesson: we used to call the boned and rolled loin a 'loin fillet' and we seldom sold them. Then one day a caterer friend commented, 'These fillets are all the rage this year on *Masterchef*. They're all using them but they call them 'cannons'.' Who were we to argue with *Masterchef*? And the name sounded much more expensive! We changed the name on the labels and put a board out advertising 'Cannon of lamb' and sold out most weeks from then onwards. That board was a good selling tool as it generated interest and many people who came to ask what a cannon was ended up buying one.

There were days when I wondered whether doing our own butchery was sensible. It took at least a day and often more of my time each week and it was especially difficult during lambing when I could be moving back and forth between the lambing shed and butchery (with the consequent endless changes of clothes and hand washing) several times a day.

But it had its benefits. It meant I could adapt the butchery of each animal to suit the market we were attending and accommodate the inevitable variation in condition and fatness of the lambs. And it was nice to have a day doing something different out of the wet and mud in the winter or the sweat and dust in the summer. It was also a day that Frances and I got to work together and we could chat or listen to the radio. We had a well-established routine which even the dog had become part of – he seemed to know exactly when I'd get to the first bit of trim – and he'd be waiting outside the door.

We got quite good at predicting demand at the markets. Hereford always wanted deals and discounts, Ludlow wanted the opposite and we could always sell cannons there. Leominster was a small but very profitable market with very loyal customers who valued good quality and needed to be looked after. Leominster customers were also very good at ordering in advance which was the holy grail of market trading as it took the guesswork out of the preparation.

We had our first-ever on-line order for a Leominster customer, a delightful elderly lady called Mrs Oxford who ordered a small pack of Lamb's Liver. As I printed this out to take to the packing room, it occurred to me that the cost of the paper and ink was probably just about equivalent to the profit on the order! We sold a lot by recommendation at Leominster and it was always a very enjoyable Saturday morning among friends once a month.

The downside of doing our own butchery was that it depended on me being available every week. I didn't mind doing it – in fact I quite enjoyed it – but it took me away from other tasks and there was always the risk that I might get sick and be unable to do it. So more for security than workload, we decided to look for a 'backup butcher' to help out when necessary.

We had a procession of them... 'I've been doing this for forty years,' was their attitude with the implication that I couldn't teach them anything. In reality they were used to cutting for a butcher's slab and couldn't get their heads around the need to cut for display on a market stall.

Ray was the classic butcher, huge arms with a jovial approach to life and unfailing self-confidence. He was also a religious chap. 'You do realise that a lamb carcase and a human skeleton are so similar that it disproves the theory of evolution?' was one of his opening lines. His butchery was about as good as his theology, so he didn't last long.

After several more false starts, we found Barry, a pleasant chap with a nice line in (secular) chatter and pretty good butchery skills. But best of all he was adaptable and willing to work to our standards. In fact he made some useful suggestions as we worked away together, usually under pressure to get the prep done for a market, and he proved to be a great help in busy periods, especially during lambing.

I dashed over from the lambing shed one morning as he arrived. 'Just five to do today,' I said breezily. 'Come into the cutting room and I'll explain what I need. There's two markets to cut for and mince and diced mutton to prepare.' Barry looked at me quizzically as I rattled through what I wanted done.

With a resigned sigh he mumbled, 'OK – there's a lot to do here so I'd best get on.'

I left him to it and went back to the shed. I saw him briefly at lunch time and he looked harassed but carried on. We then got really busy and before I realised, it was seven o'clock. As things calmed down, I decided to pop back to the house for some food but as I opened the door I saw a figure stomping up the slope into the yard. It was Barry and he'd only just finished.

I couldn't see his face in the gloom but I sensed he wasn't happy so I closed the door quietly until he'd gone. I'm not sure the older butchers would have done that, they'd have just left us to cope. So Barry got a bit extra in his pay that week.

THINGS DIDN'T ALWAYS go right for us. We had our fair share of disasters, some natural and unpredictable and others rather more self-inflicted. I've chronicled some of our wrong turnings in this book, in the hope that others can avoid them or at least have a quiet smile to themselves.

Running any business is all about making decisions and the general idea is to make more good ones than bad ones. Wrong decisions are a bit like turnings on a journey: they can be like driving into a blind alley which is annoying but easily rectified by turning around and driving out again. Or they can be like crossing a bridge and finding yourself on the wrong side of a river with no way back until you find another bridge.

Our first but by no means our worst 'bridge-crossing' decision was a change in ewe breed. Our initial plan had been to do what everyone else did – use commercial, crossbred ewes with a Texel ram to produce fat lambs. This worked well but had two drawbacks, the first being that we had to buy in our replacement ewes with the consequent risk of importing disease. The second was that it wasn't very 'local' – the ewes generally came through the local market but they had varied breeding and it was difficult to get any sort of provenance which was important to us and our customers.

One of our neighbours was experimenting with the Lleyn breed at the time and they seemed to offer a solution. They had been taken on by the organic fraternity as an ideal sheep for low input systems and had become very popular in our area. Our idea was to breed them as a self-contained flock to avoid the need to buy in replacement ewes.

'As ever, if we'd stayed with our original plan, things would have been fine.'

This is how we got to know Edward Collins, our local Lleyn breeder, who persuaded me to buy ewe lambs and lamb them as lambs, i.e. at twelve months old. 'You don't need yearlings,' he argued, probably because he didn't have any left to sell. 'Ewe lambs are cheaper and you'll get an extra crop of lambs from them. Just make sure they only rear one lamb in their first year and put the second one from any doubles on a feeder. We do it every year – it's easy!'

With hindsight, this wasn't the best idea for relatively novice sheep-keepers like us, but we made it work. We bought forty pedigree Lleyn ewe lambs and two good rams from him and set about developing our new flock. Lambing went well with this group; they were in great condition and we looked after them carefully.

As ever, if we'd stayed with the original plan, things would have been fine, but I also bought thirteen Lleyn yearlings from a friend. Unfortunately, these were the so-called 'unimproved' type which were smaller and didn't compete well with our big mule crosses. They also suffered from the old Lleyn problem of being over-prolific and produced far too many multiple (more than two) births. We had a very difficult time with them and struggled with low birthweights and poor milk supply.

Despite this minor setback, the lambs grew well and we were happy with our new flock, but as the season progressed we ran into another problem. These pure bred Lleyn lambs grew quickly all right but either finished at too low a weight or got too fat. The only way to manage this was to rear them separately from the rest.

I decided to persevere into a second season as we had invested too much to change back again. We weren't so much on the wrong side of the river as in the middle of it, unable either to afford to change over completely or to cut our losses and go back to our original crossbreds.

We managed the flock more carefully that winter, feeding the Lleyns in a separate group so they didn't get pushed out by the bigger mules and we had a much better season – although still too many multiple births. But it was all just too complicated and the need to rear the lambs separately to stop them getting over-fat was the final death knell to our plan. It wasn't so disastrous that we needed to sell the Lleyns we already had but we decided to revert to a more conventional crossbreed going forward.

I know I risk the wrath of Lleyn enthusiasts with these comments and I am sure that as a breed, they are fine – especially the improved ones – but they didn't suit our system. We couldn't afford a complete changeover in one go, so we were left with trying to manage a mixed flock and it didn't work.

This was our first setback; not serious but a timely reminder that we were in this for real now without earnings from employment to rescue us if things went wrong. We'd learned a lot in those early full-time years; how to rid ourselves of the hobby mentality, how to grow the business with more sheep and more land, and how to prepare and sell our produce to a growing band of customers. We'd learned about butchery and the folly of trying to save money with DIY facilities. We'd learned the value of good relationships and the consequences of taking advice from others who don't always have your interests at heart.

But most of all we'd learned that we enjoyed doing this and wanted to carry on. We were to meet many more setbacks, along the way. But as we became more battle-hardened and streetwise, it became clear that all we really needed, apart from limitless cash, thirty-hour days and eight-day weeks, was an unfailing sense of humour.

4. Could Do Better

IT WAS TIME TO TAKE stock. We'd sailed through our hobby years, full of excitement and optimism; we were (or at least I was) doing what we'd always wanted to do. Our early full-time years had been more challenging but were still mostly fun and we'd learned a lot. We'd reached about a hundred ewes on forty acres with a decent barn to lamb them in. We had a fledgling retail business and the facilities to grow it and, more importantly, an increasing and enthusiastic band of customers who enjoyed our lamb and kept coming back for more. We'd spent a lot of money on a cold room and butchery, on more sheep and on more rent and we were selling everything we could produce. But so far our income stubbornly refused to exceed our outgoings and we were losing money.

I had a very heated debate one morning at the abattoir when I suggested that our lambs cost about a £100 each to produce.

'Don't talk daft,' was Alf the head slaughterman's verdict. 'A bit of grub, some worm drench and our killing charge. Ten to fifteen quid tops!'

'So what about the new clutch in the Land Rover,' I retorted. 'That's a pound a lamb this year before we start!'

'Well if you're going to load up those costs, what do you expect!'

'I wouldn't have the Land Rover if I didn't have the sheep – it's a legitimate farm expense that has to be paid for!'

Even our customers didn't understand. 'I don't see why your lamb costs so much,' argued Ben, one of our early commercial customers. 'You've no costs, have you? All they eat is grass.' I mumbled about overheads and rent but he wasn't listening.

And so it went on. Most people, even sometimes those involved in the industry like Alf, have no idea of the costs involved in farming sheep. Even farmers sometimes fool themselves by 'losing' costs in other parts of the business, often using other staff to help out at lambing or using machinery that's paid for by the arable crops. We didn't have that option; we only had the sheep so all the costs were down to them – and they were too high.

Typically, the average cost of producing a lamb in the UK lowlands was about £80 (2012 prices) and since most farmers sold their fat lambs for about the same figure, it's not hard to see why sheep farming rarely made any money. In our case, we were selling our sheep for at least twice that price but our high costs kept our bank account firmly in the red.

I often used the excuse that, as a small operation, our costs were bound to be higher, but the reality was stark and undeniable: we were inefficient and our next step had to be to try and fix that.

IN A VAIN ATTEMPT to think strategically, I approached this quest for efficiency with some logic. We could spend less and produce the same or we could produce more from the same outlay. Or of course, in an ideal world, we could do both at once. Some of these changes related to the sheep themselves and some to our equipment and how we did things.

An early change we tried was to mix our own concentrates – what farmers call 'hard feed' to distinguish it from hay and silage, which is usually called 'forage'. Pre-formulated feeds are expensive and always a bit of a compromise for small farmers like us who can't afford to buy different formulations for different classes of stock. The idea was that I could formulate a mix of cereals (oats bought from a neighbour) and purchased protein and mineral supplements for each group and prepare it in the quantities we needed. Each group would get exactly what they needed so we'd be more efficient and we'd save money. It was also a bit more 'local' than buying in from a national feed supplier.

All was fine at first when, full of enthusiasm, I pinned my 'recipe' sheet for my carefully calculated rations on the grain bin. But the enthusiasm was short-lived. It was back-breaking work and it took a while to do it properly. The choking dust swirling around didn't help either, and once we started lambing, I simply didn't have the time or the energy to do it properly. I'd also overestimated the cooperation I'd get from the sheep who were determined not to eat the important bits like the protein pellets and scoffed the oats down at such a rate that they ended up choking. I had several of them foaming at the mouth one day and very nearly lost one of them who'd eaten so quickly she'd blocked her throat. I'm firmly of the view that sheep aren't stupid but they were clearly trying to convince me otherwise that morning. So we didn't reduce waste and it really wasn't that much cheaper – the oats were but the protein and mineral pellets, in the volumes we were buying, were too expensive. So overall we were no better off financially and I was considerably more dusty and had an aching back.

We also tried to improve our pre-lambing feeding. The traditional practice of feeding an increasing quantity of concentrates in the six weeks leading up to lambing was now frowned upon as it favoured the unborn lambs but didn't really benefit the ewe. A better idea was so called 'flat rate' feeding, where we fed the same overall quantity but spread it evenly over the six weeks so that the ewe got her share in the first three weeks while the lambs were tiny and they in turn took theirs as they grew to full term. Using the same amount of feed didn't save us money in the short term but would ensure we had ewes lambing in good condition with plenty of milk, and lambs that were a good size without being too big, that would grow away well. All being well, this would mean we had lambs ready earlier and with less need for additional feed later in the year – hopefully a welfare and financial bonus.

Sadly, we chose the wrong year to do this. Our ewes came inside in better condition than ever before, and feeding concentrates early allowed them to lay down too much abdominal fat so that as their lambs grew, they became very prone to prolapsing. This is a very unpleasant condition, much debated as to its cause, but it is principally due to pressure inside the abdomen which causes the ewe to think she is lambing and push some or – in catastrophic cases – most of her uterus out. If we caught them early enough, we could put minor cases back in and hold them in place with a harness. It took strength, patience and lubrication but was usually successful. But the bad ones were a vet's job and were usually terminal, with the loss of the ewe and her unborn lambs. That year we lost four ewes and their eight lambs in the run-up to lambing, and had another eight or so with harnesses on. A terrible welfare cost, distressing to us and the sheep, and equally bad for the bank balance. We should have regularly 'condition-scored' the ewes, which meant catching them and feeling the flesh cover on their backs so that we could adjust the feeding as we went along. But we simply didn't have the time in our busy schedule – an early indicator of future tension between good farming practice and a demanding retailing business.

ONE OF OUR BIGGEST problems was that we were constrained by biology. Sheep are seasonal breeders, their lambs are born in the spring, mature through the summer and are ready for market in the autumn. We could vary lambing dates a little, by putting the rams in earlier or later, but only by a month or two. All pretty obvious really but a problem if you need a steady flow of lambs all year for your retail business. Trying to produce lambs out of season was costing money in extra bought-in feed and in housing during the winter months when we had no access to grazing.

My plan was to produce lambs that would grow at different rates and be ready at different times. All quite simple really.

I tried to explain this to Frances when we were planning our tupping dates (the date the rams are put in with the ewes) for the coming season. As usual, she thought I was making life too complicated.

'If we lamb a group in early March ahead of the main flock, we stand a chance of getting some of the biggest lambs ready to go by June,' was my first attempt.

'But it's cold then and the grass won't be growing,' she replied, ever the pragmatist.

I tried again. 'I also think we should try lambing ewe lambs, like Ed Collins showed us. We can get the numbers up quicker, and if we lamb them later as he does, we'll have a group that'll be ready in the late winter/early spring the following year.'

'But it's a lot of work. You're always worn out after lambing the main flock – why put yourself through it two or even three times? It also means we'll have bunches of lambs that need all the treatments and weighing separately from the main group, and that'll be more work. Why don't we just wait and see? Aren't they like humans – some lambs grow fast and some don't. Some get fat quickly and some don't?'

But I was hung up about it, so we put in place these new routines to try to even out the lamb flow.

And of course Frances was right. All this complication wasn't necessary, as the lambs naturally grew at different rates and we found the best option was to weigh them regularly and keep them in just three separate groups – those that would be ready in the next few weeks and would 'finish' on grass alone, the 'medium keep' lambs that would be ready later in the winter and would grow on decent silage with a little concentrate feed, and 'long keep' ones that could be outwintered on some sheltered winter grazing that we rented just up the road. So we put everything back where it was. We abandoned lambing ewe lambs and settled on a lambing date at the end of March. We were slowly beginning to learn the most important lesson of all; that simple stuff gets done properly.

Another change we made was the way we treated foot rot problems in the flock. This was about saving the 'hidden' cost of poor welfare and its effect on productivity. In other words, sheep with bad feet don't do well. They don't graze as effectively and lose condition. Thinner ewes are then more difficult to get pregnant and tend to have a higher proportion of single lambs. They also have less milk when they do lamb, and hence their lambs don't thrive.

We'd stopped using traditional foot bathing as it was just too hard. We either had to bring the sheep home or take the footbath to them. They usually refused to walk through it and we had to resort to spreading straw over it or virtually lifting and carrying each sheep and standing her in the bath. We finally gave up when one especially uncooperative ewe decided to kick her back feet up as she left the bath, spraying the formalin solution into my eyes, leaving me howling with pain and looking for a tap to douse my face. Not only was it hard and hazardous, foot-bathing the way we were doing it wasn't very effective either.

A little aside here: we did try foot-bathing again many years later when we had a batch of lambs to sell at market and, as sheep do, they started to limp and hold their feet up a couple of weeks before the sale was due. We did the job properly this time and set up the footbath in the barn with the exit facing the light through open doors. We fashioned a drying-off pen on concrete so they could stand there while the solution dried on their feet and we did it every day for a week. The animals were at home so it was easy to do and it worked like a dream. Every afternoon we brought them inside, they queued up in a nice orderly line, walked serenely through the bath towards the light and stood patiently while their feet dried. And after a few days not one was limping, we were relaxed and had no bruised arms or damaged backs from wrestling sheep – and my eyes were pleasantly free from formalin.

But back to the present. We'd decided that the easiest way to treat foot rot was to catch individual animals and treat them with antibiotic spray and injections, what the vets call 'topical treatment'. This worked fine while they were outside, but less so when we brought them inside for lambing. The warm bedding was the ideal breeding ground for the foot rot bugs, and if one animal had it, it would spread to the others within days. So Harriet and I decided to vaccinate them as they came in. I persuaded her to come and help me one afternoon and we did the whole flock in a couple of hours. Unfortunately,

'The big advantage of silage was we could cut it earlier.'

her definition of 'help' was that she'd do the vaccinating while I did the catching, which led to a very tiring afternoon. I was discovering that most sheep-related jobs involve work with the potential to do damage to your back, but in this instance, it was worth it and we had only one foot rot case that winter. We also had the unexpected bonus of much reduced 'scald' (a sort of junior foot rot) in the lambs the following spring.

PERHAPS THE MOST IMPORTANT change we made to our farming routine was to switch from hay to silage.

Making hay has been a farm activity since biblical times. Grass grows rapidly in the summer and very little or not at all in the winter. The idea is that surplus grass in the summer can be cut and dried in the sun and then stored for use in the winter. It's been done this way for centuries, but there are problems, the obvious one being that you need sunshine – typically four or five days of it, ideally with a nice breeze, to dry the hay. So haymaking is always a race against the weather and crops can be ruined by unexpected downpours.

The second problem is that grass needs to be cut late in the

'The equipment needed is way out of reach financially for all but the largest farms.'

season when it's quite 'stemmy' or it can be very difficult to dry. Late-cut grass has a lower feed value as it contains more stem and less leaf and therefore needs more supplementation with bought-in feed. And finally, cutting the grass late in the season means that the fields aren't available for grazing again until the autumn.

The principle with silage-making was the same; we still cut the surplus summer grass and kept it for use in the winter, but with silage we stored it damp, in bales wrapped in plastic film. The plastic kept the air out and the moisture allowed the sugars in the grass to ferment to produce acetic acid or vinegar, which conserved the grass. Effectively it was pickled grass. The big advantage of silage was that we could cut it earlier in the year, as it didn't need to dry. This meant it was more leafy and a better feed, and cutting earlier meant we could get back to grazing the fields sooner – with a bit of luck at the point we weaned the lambs. The other advantage was that the plastic-wrapped bales could be stored outside and in the places where they were needed which was especially useful for sheep wintered away from home.

The downside of silage making was that, for a small business like ours, it was a contractor's job and therefore costly. The equipment needed was way out of reach financially for all but the largest farms.

The move from hay to silage had been a well-trodden path on livestock farms since my early ADAS days and for us, despite the extra costs and the care required, the move was a game changer. The better quality forage meant that we more than covered the extra costs through savings in bought-in feed. Storing it outside meant we had more room in the shed for more ewes as our flock grew, and the fast and efficient contractor job meant we had the fields back growing grass more quickly, ready for the young lambs as they were weaned.

As with shearers and scanners, our experience with silage contractors was that if you are fair with them, they will reciprocate. I never cut grass until I had agreed a date for baling and wrapping with Kevin (another Kevin) and he knew I wouldn't call him at short notice just as it was about to rain, as many of his other customers did.

Our change to silage wasn't completely trouble-free though. In our first year, we carelessly allowed the airtight seal of the plastic wrap to get damaged by rough handling and also by birds. The air got in and little mushroom-like growths started to appear, growing through the plastic as the mould developed underneath. It was a very dispiriting job unwrapping bales in the winter and finding huge patches of white mould on them which we had to spend hours laboriously pulling out before the sheep got near it. We were assured that it would be listeria mould and that the sheep would die if they ate it. That first year was so bad that I eventually unwrapped each bale outside to make moving the mouldy stuff to the muck heap easier.

Our non-farming friends were aghast when we roped them

'I moved the bales as soon as they were wrapped to the side of the field and covered them with a net.'

in to help one weekend and implored us to go back to hay, but we persevered. The next year, we used the recommended six layers of wrap instead of the four preferred by parsimonious Herefordshire farmers, and I moved the bales as soon as they were wrapped to the side of the field and covered them with a net. These nets kept the birds off and meant I could then cart the bales home when I had enough time to do it carefully with carpets on the trailer and even bed sheets on the thorn bushes in some field gateways to avoid damaging the plastic.

Our other precaution was to get 'Eddie the Mole Man' in to remove the moles from our silage fields, as we were warned that soil from mole hills was a major source of listeria contamination. This was the source of the oft-quoted saying from Frances that 'Moles kill sheep!' I'm sure some of our visitors had visions of a crowd of moles mercilessly chasing and dragging down a ewe, but it was a genuine problem, and Eddie's charges were yet another cost that Alf the abattoir guy had chosen to overlook.

I have been laughed at many times for the care I took over

silage-making after that difficult first year, but I was vindicated some years later when I had to buy in a few extra bales from a neighbour to see us through a long winter. I took at least a barrow load of waste out of each bought-in bale – compared to two barrow loads from the entire stack of a hundred bales of our own stuff. Enough said.

The only other downsides were the knock to our monthly cashflow every year when the contractor's bill came in and the fact that using a contractor meant I was moving further away from the practical farming that I loved. But of course it released my time for more important things.

AT FIRST SIGHT, ERECTING another shed to save money seemed rather counter-intuitive. But as Stan the barn erector had predicted, we very quickly ran out of room in the new barn and it was a significant restriction on our flock size, especially in the run-up to lambing. We were committed to lambing inside as we didn't have access to winter grass and our limit in the main barn was about one hundred and fifty ewes. We really needed enough space for another fifty ewes at least to make efficient use of our time at lambing and to give us more lambs to sell. So adding an extension to the existing barn was more about removing a constraint to growth than saving cash.

I priced up a lean-to from Browns Of Wem, the original barn supplier, to run the length of the existing one, but the cost was about the same as the original building. So, true to form, we decided to build one ourselves. We built a pole barn – which is not as some have suggested, a structure built by our East European friends, but one built with second-hand timber; in our case, with redundant hop poles. Kevin and I went off one evening to fetch them – my loader tractor to lift them and his, with a massive trailer on, to carry them home. They were huge and I just could not see how we would get them upright.

'At first sight, erecting another shed to save money seemed rather counter-intuitive.'

The following weekend Kevin arrived with an ancient post hole borer to go on my tractor and we managed to bore fifteen six-foot-deep holes to put the poles in. We designed it as we went, but it worked out pretty well – especially the 'crane jib' which we made for my tractor loader to help us lift the horizontal poles into place.

We added a couple of bays beyond the end of the existing barn so we could stack bales in easily with the tractor, and although we bought new cross-timbers and roofing sheets, we still managed to put up a building measuring one hundred and ten feet by fifteen feet for just over £2,000. Although an extra immediate cost, this enabled us to lamb more ewes with the same staff and added significantly to our bottom line.

The final episode in this catalogue of changes made to save time and money was when I smashed a hole in the wall between the main barn and the new lean-to. Animals and people could now move from one to the other without going outside. The day I spent making that hole, including cutting out the reinforcing steel that Frances and I had laboriously installed some years earlier, and the second I invested in making a door and hanging it, were probably two of the most productive days in my entire farming life. Suddenly we could lamb out in the lean-to and bring ewes and lambs into the pens in the main barn without the usual rodeo. We could move equipment and people around easily and, most important, I could go to check on ewes out there on a wet night without having to find a torch and put my coat on. Probably the most impressive and certainly the most effective improvement I made and such an obvious thing, one wonders why I didn't think of it sooner...

We were making progress; we weren't working less hard, but we were managing a much bigger business. Our Achilles heel, however, was still that our land was so spread out and we spent far too much time travelling around or moving animals from one place to another.

This was a problem that we couldn't easily fix. What we really needed was a big block of land to rent – maybe even a farm. But that would have to wait.

IT WAS DAWNING ON US that while doing the farming well was important, for our reputation and peace of mind as well as for our business, it was the retailing that would really make or break us. However good the farming was, if we didn't sell the product successfully, we'd have no income and the farming would go under.

I was hugely excited about our early success at farmers' markets. 'I think we should go to as many as we can,' I enthused. 'The more we attend, the more we'll sell. We need to get ourselves and our product in front of as many people as possible. The farming is about controlling costs, whereas selling is about throughput. The only extra costs are the stall fees and a bit of marketing stuff. It's got to be worth doing,' I concluded perhaps a little over enthusiastically.

'And who's going to attend all these markets?' asked Frances. 'Each one is a day off the farm and there's all the prep to do as well.'

'But prepping for more markets won't take that much longer,' I argued, 'and if we get a second chiller, we could even do more than one market a day.'

'With both of us away from the farm, working for nothing.'

Frances wasn't keen as she knew that I didn't count the cost of our time in my calculations. There isn't an artisan food business in the country that would be profitable if the value of the time invested by owners and families was included as a cost. A chastening thought but very true. But ever supportive, she agreed, and we proceeded to add markets to our portfolio until we were attending thirteen separate events every month.

Financially, the markets had made it possible for us to move from a busy hobby to a full-time producer/retailer business. At our peak, we were taking about sixty percent of our meat sales income from market trading, and there was a significant spin-off from direct sales from the farm as customers got to know us.

As I hinted in the last chapter, these markets were all very different, ranging from the very busy Ludlow event to the smaller ones held in pub carparks and even one in a vicarage garden! They were great fun; the banter with the other traders was always a laugh and the customers were an absolute delight. But we had to learn some hard lessons quickly, not least that we had to sell what customers wanted to buy, in volumes that would cover our costs.

Caroline turned up in her gleaming Jaguar one cold, wet Saturday morning in Hereford. We were setting up the Farmers' Market ready for our monthly attempt to part the public

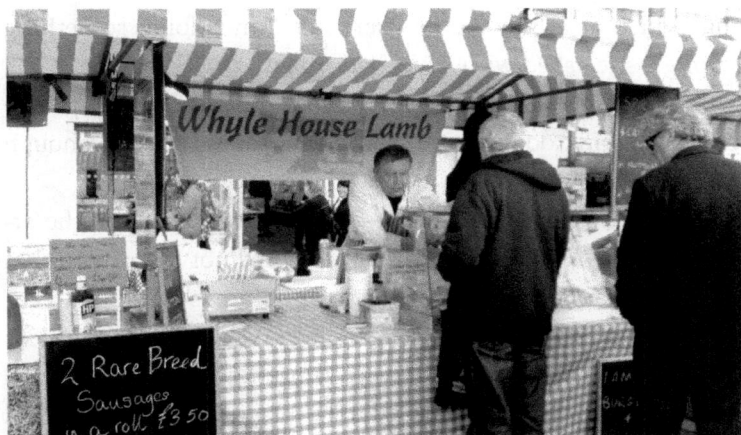

'I was hugely excited about our early success at farmers' markets.'

from their cash and she was late, which didn't endear her to the rest of us. We'd been there since seven in the driving sleet setting up the stalls, and she turned up just as we had finished. Her designer clothes and expensive wellies contrasted with the rest of us in crumpled but warm market clothes and boots. Her plan was to sell quality coffee, beautifully packaged and presented to the good folk of Hereford because she thought they 'deserved it.' I'm pretty sure she didn't sell anything that day.

The good folk of Hereford failed to see that they needed her coffee. Perhaps if she'd been selling other produce that they did need, she could have encouraged them to try her coffee, but as it was, she was trying to sell what she thought they should have rather than what they wanted to buy.

Sarah was another one – she made cakes. Excellent cakes they were too, made with care and passion and with lots of love. Clearly people want – or even 'need'– cake, but care and passion will only carry you so far. On this particular day, as one of the regular traders, I had been assigned to help her and I stood quietly by her stall studying her products and pricing, which clearly made her nervous. 'Am I doing something wrong?' she asked with a worried frown.

'No,' I replied, 'but even if you sell everything you've brought along today, which is unlikely, you won't take enough to cover the stall fee.'

There are lots of people like Caroline and Sarah in the artisan food world, and we were determined not to fall into that trap. We planned our pricing and our volumes carefully to ensure we had a chance of making money. We were quite ruthless about stopping markets that were unprofitable and we quizzed market managers of ones we hoped to attend about footfall and competition. But above all, we tried to produce what we knew we could sell, and we always took the view that if we made a loss it was our fault, not the customer's.

A Twitter (X) acquaintance of mine was complaining recently that he'd had to give up 'pasture-fed' beef production because people wouldn't pay enough for it. While I'm sorry that his dream to create what he believes is a 'better' product hasn't been successful, I do wonder why we blame the customer when a business we create doesn't work. It's not up to others to fund our unprofitable business choices. It's up to us to set up a business in such a way that it at least has a chance of showing a profit. Of course there are plenty of challenges that will drive a business plan off course, but we should at least start with a plan that stacks up financially.

And this idealism isn't limited to farmers' market traders. The current hot topic in the farming world is 'regenerative agriculture' and like all 'new' initiatives, it isn't new. Nobody can agree exactly what it means, consultants are making a lot of money devising blueprints for the disciples to follow, and the advocates are strangely reticent to disclose the economic consequences of adopting it. Having said that, 'regen ag' as it's known to its friends, contains a lot of useful ideas about soil health and biology, and particularly about carbon capture, but it's in danger of being lost under a morass of proselytising zeal. It reminds me of the start of the organic movement in the sixties when genuine and laudable ideas were drowned out by the rhetoric and it's taken several decades for that thinking to become mainstream.

It's important that we as an industry are innovative; we need to produce food to high welfare standards and with a lower environmental footprint, but it has to be financially sustainable. And that means that we can't rewrite the basic rules of business; that the value of the product to the consumer – what they'll pay for it – has to be greater than the cost of producing it. Some will pay a premium for products with an 'environmental' or 'welfare' label but, sadly, in the UK most won't. Even the organic movement which has been around for decades has yet to gain more than a couple of percent of the total food and drink market in the UK. A recent AHDB report stated that while the consumers are 'willing to buy organic, price remains a key barrier', and gave the organic share of total food and drink volume as 1.2%. Some producers claim it's a matter of educating their customers to understand why they should buy their products; others are content to supply a niche and largely unexpandable market; and still others call for a wholesale change in consumer attitudes to food pricing. We figured we'd go bust while we were trying to do any of these things and decided we'd be better off taking a pragmatic approach.

So we took the simple view that we should sell products that customers understood, that they believed they needed, and that they could afford. To that end, we indulged in what might have seemed a lot of good natured chatter on the stall but was actually an exercise in gathering intelligence.

In the early days, we were often asked for chump chops

which was a cut we didn't do. They were hard to butcher neatly and we realised quickly that if we called the 'chump' by its other name – the 'rump' – and boned and rolled it, we could sell it as a small 'midweek roasting joint'. We also had debates about Barnsley chops – double loin chops – mostly with people from that part of Yorkshire where they are called 'butterfly' chops. And for awhile we sold 'value chops' which were tiny neck chops. Being cheap, they sold well in Hereford.

It was all harmless fun and a good way to engage with our customers, but we did have one serious debate that changed how we distinguished between 'fresh' and 'frozen' meat. 'Does that mean the frozen stuff isn't fresh then?' quipped one of our regulars. From that day forth, the opposite to 'frozen' was always 'chilled'.

AS LONG AS THE WEATHER played along, markets were a very attractive way to spend a few hours away from the farm. Of course, bad weather could damage sales, but once we became regular traders on a market, it was amazing how people would turn out to support us even on the wettest of days. From a trader point of view though, the real weather problem was wind, as we found out one Christmas eve.

It was the last market of the year. We all had lots of Christmas orders for customers to collect or I'm sure we would have given up much sooner. The wind was howling through High Town in Hereford and some of the less enthusiastic traders had packed up and gone home early, leaving their stalls open. The wind got under one of these empty stalls, lifted the heavy eight-by-four-foot table off the ground and blew it like a frisbee across the square – narrowly missing a small child. The child broke down in hysterics, but luckily its mother was of the 'you-weren't-hurt-so-stop-making-a-fuss' school. We managed to calm down the child and recover the board without further

'You can have your lamb chops delivered free by helicopter, just £100 each.'

incident. That truly memorable market finally came to an end when Enid, our veg-selling friend, was quite literally blown clean off her feet!

Most markets weren't that challenging but even quite minor breezes could blow materials off the stall. Stronger winds caused the roofs and the back sheets to flap and, in some cases, become airborne. We became very adept at dealing with bad weather but a nice, still day was always a welcome sight when arriving on a market square early in the morning.

TO HELP IMPROVE our visibility at markets, we decided we needed to improve our brand image. I wasn't sure what a brand image was but I figured we needed to have one and do it well. 'You can have your lamb chops delivered free by helicopter, just £100 each,' was one of our sillier promotions. My son, Ross, had a helicopter licence and he would occasionally fly over with friends and land on our field, usually when we were lambing. On this occasion, his friend's parents had ordered some lamb chops, so he took them back with him. It made a

*'Grandson Ollie provided a
valuable service...'*

great photo op but at £400 per hour flying time, it wasn't exactly cost-effective.

We tried hard to be professional with a consistent colour scheme and easily recognisable labels so our customers never struggled to find us, even at the busiest markets.

Much of our marketing was aimed at 'Emma'. She was an entirely fictitious customer whose description ran to several pages but in essence was a young mum, with a busy life, young kids, a reasonable income and a desire and interest to eat healthily and feed her family well. She was interested but not that knowledgeable about food and was time-poor. This image was really helpful when designing promotional material and when thinking about new products – like a pack of diced lamb of exactly the right weight to go with a bottle of artisan cook-in tagine sauce. Our friends bought into the

'It's an amazing event when more than sixteen hundred farmers open their doors for one Sunday each year.'

idea too and would excitedly call us with, 'We've found you another Emma!'

We wanted our customers to engage with us – I'd heard marketing people use the word 'engage' so we figured it was important. Our idea was to create a 'Customer Family' and involve them in our business. Open Farm Sunday, organised by a wonderful group called LEAF (Linking the Environment and Farming), was our first attempt. It's an amazing event when more than sixteen hundred farmers open their doors for one Sunday each year. Visitor numbers topped two point seven million pre-pandemic and are growing again now. We had two very successful years, aided by family helpers. With tractor and trailer rides by Ross and talks by me along the route, butchery talks by Hannah, a BBQ by Kate and Alex, and food and drinks from the local church ladies. Grandson Ollie provided a valuable service, demonstrating to other four year olds how to open the sheep race gates, walk through and weigh themselves over and over again.

The problem with Open Farm Sunday was that it took a

lot of time to set up and always fell when we were busy trying to make silage. It did however convince us that attracting the public to our farm was a good idea and that's how our Lambing Afternoons started.

'Come and see a lamb born, help with the chores and bottle-feed the pet lambs' was a pretty compelling draw and we involved the local radio and press who ran stories for us. Our farming friends said we were mad. 'What are you going to do with the dead ones?' was their initial encouraging contribution. And funnily enough this did happen in our first year. Just as a family group arrived with several small children, a lamb I was attempting to revive decided to end its days and one of my young helpers had to smuggle it out of the barn under his coat! But these afternoons were a great success and we were inundated.

For the first two years, we opened our doors every day for three weeks, from two to five in the afternoon, but this became a real chore. It meant one of us had to be in the shed all the

*'These afternoons were a great success
and we were inundated.'*

"Come and see a lamb born, help with the chores and bottle-feed the pet lambs' was a pretty compelling draw..'

time and it severely restricted what other jobs we could do, so after that it became weekends only.

Health and Safety and Risk Assessments were diligently prepared, and never referred to again. As far as we could see, there were only two serious issues to worry about. The first was the tendency of small children to put their fingers in their mouths; the second was the risk to pregnant women from the virus that causes sheep to abort.

The finger-sucking was simply about hygiene which most parents seemed to ignore. We encouraged the children to wash their hands and gave them hand sanitiser, and as far as we know, no one ever came to any harm.

The second issue was more difficult.

'I don't think I'll come in after all,' muttered the young woman with an ashen face. 'I'm a vegetarian, I think I'll give it a miss and wait in the car.' I used to welcome parties at the shed door and explain the rules to keep them and the animals safe and I always finished with, 'And if you're pregnant you can't come in.'

My talks were quite forceful but didn't usually have that effect. I wandered over to her car to check on her later in the afternoon and she confided she was pregnant but hadn't yet told her husband! Frances had warned me this would happen one day, and after this incident she banned me from mentioning it in the welcome talk. We also crafted a clearer notice explaining the risks and stuck it firmly on the lambing shed door.

Lambing Afternoons were a special time for us and I never got tired of answering our visitors' questions – even though they were nearly always the same. 'How long do you keep the ewes?' 'How old are the lambs when they go for meat?' 'What breed are they?' Our friends Tim and Stacey were a great help on these occasions as they'd heard it all before and could look after groups of visitors if we were busy. We never charged for the sessions but we had a board by the door suggesting that they might buy some lamb if they'd enjoyed their visit. This was a source of incredulity to family and friends. 'You'll never sell lamb chops to people who've just fed baby lambs on a bottle,' was the view, but we did – often £20 or £30 worth – and visitors came back year after year.

One of the nicest parts about Lambing Afternoons were the regular visits by the local Rainbows (baby Brownies) group. They came every year and fed the pet lambs, walked up the fields to feed the ewes and marvelled at the tiny new-borns, but they never actually saw a lamb come into the world – until one year, just before they arrived, a ewe bearing a single lamb went into labour. Nothing happened for a while so we did all the other jobs and I then suggested we'd have a look to see if everything was OK. This quick look turned into a serious problem with a huge lamb that needed help and despite being warned that it might not end well, they all stayed, absolutely riveted. Thirty minutes later I produced a live lamb to applause from the girls and the odd wince from their mums.

THINGS WERE LOOKING UP. We were selling well at most of our markets and were developing a decent direct trade. Our reputation was growing, as was a band of enthusiastic customers. We were getting better at the farming too, but we just couldn't produce enough lamb from our own flock. Our farming and our retail business had got out of step and, try as we might, we just couldn't get the numbers up. I'd go off to Ludlow market each autumn and buy more ewes, but then we'd find with dismay that we'd have to cull a similar number – for foot problems or mastitis. So the overall flock size would stay almost the same. Without significant investment (which we didn't have) it was going to take a long time to build our flock numbers in line with demand. We just had to find a way of producing more lambs without spending too much money.

So I had an idea. Not one of my better ones as it turned out but even my harshest critics couldn't deny the logic. Instead of buying more ewes to produce more lambs, why not cut out the expensive 'ewe' step and just buy the lambs 'ready born'? All sheep farmers have spare lambs at lambing time. They're called 'tiddlers' in Herefordshire, and we knew a good local chap who would sell us his to rear. We already had the milk feeders to rear our own orphan lambs and third triplets, so this would be a simple extension of what we were already doing.

So, what could possibly go wrong? As it turned out, quite a lot.

Once in Royal David's City on the car radio drifted across the Hunt Kennels car park as I loaded dead lambs into the skip. It was Christmas Eve and the mist hung over the dank and dreary scene. I used to get so excited on Christmas eve as a kid – Dad would bring in the tree and we'd decorate it with exactly the same baubles and a fairy every year. And even after I grew up and moved away, travelling home on Christmas eve for a few days' break held a certain thrill and anticipation.

113

But this was a whole new ball game. The kennels were depressing at the best of times with the larger dead animals like cows and pigs, dumped unceremoniously on the floor of their cutting room, and the smaller ones like sheep and lambs 'filed', as the operator delightfully put it, in the skips outside. Ross once described the place as grotesquely compelling – you don't really want to look inside but can't help yourself. Even on a cold day like this, the smell was unpleasant but at least there were fewer flies than usual. I didn't open the doors that afternoon; I just wasn't in the mood for the sight of lifeless creatures waiting to be 'recycled'. I stayed outside, unloaded my sombre cargo into the only empty skip I could find and left. The paperwork would have to wait; things couldn't even die now without a form in triplicate, and I'd pay next time in the 'honesty box', which was bolted firmly and rather ironically to the inside of the steel cutting room doors. The light was beginning to fail as I drove down the yard to the gate, the empty trailer rattling in the gloom making the whole place even more dispiriting. I smiled grimly, wondering what Mum and Dad would have made of this grisly exercise compared to those cosy, Christmas eve, family times.

I knew farming was going to be tough when we set out but this was painful – psychologically and financially. Fortunately, it turned out to be a brief interlude on our farming journey, but it really hadn't been a great year.

It had all started quite well. The first bought-in lambs arrived, peering uncertainly from their cardboard boxes. They'd had a pretty rough start to their short lives and you couldn't help but feel sorry for them. Some would bleat plaintively, usually from hunger, but most remained passive, probably wondering where their mum had gone. The contrast with our own lambs was stark. Ours would be bouncing around within an hour of birth on what Frances called their 'springs'. Tails wiggling ferociously as they suckled their mum's udder before snuggling down to sleep with full tummies, safe and secure. Some would even lie on top of their mum for extra warmth. These tiddler lambs had none of this reassurance. They were on their own with lacklustre coats, even at this young age, and a hollow look that usually meant an empty tummy. The old farming phrase 'Standing on a tanner (old sixpence)' described their demeanour perfectly. Our job was to provide food and a sense of security to help them restart their lives.

Food was the first priority, but some of them had already lost their suckling reflex and had to be coaxed into drinking by dribbling warm milk into their mouths. Most would eventually drink from a bottle, just like human babies, but the most serious cases had to be stomach tubed, just to get some nourishment inside them. Tubing is not for the squeamish as it involves passing a tube right down the lamb's throat into its stomach. I've struggled on many occasions trying to hold a lamb, balance the jug of milk and operate the tube all at once. To make matters harder, I'd usually decide to tube lambs late at night when all else had failed and my conscience wouldn't let me go to bed without at least trying to give them a chance of surviving until the morning. We had good lighting in our lambing barn but I'd try not to use it late at night once the ewes and lambs were settled, so late-night tubing was always done in the gloom and of course with my helpers already safely tucked up in bed.

Once the lambs were feeding from a bottle, we then had the patience-sapping and back-breaking task of getting them to use the artificial milk feeders. This involved kneeling in the pen and holding the lamb's mouth against the feeder teat with one hand while squeezing milk with the other until it got the idea. The pens were usually wet– imagine pouring six twelve-litre buckets of milk a day on the floor, although in this case it was passed through tiny lambs' kidneys first – and it was an uncomfortable position to hold for very long. Usually it took a couple of sessions, but occasionally lambs just didn't get it. I'm afraid my patience used to run out very quickly; we were trying to keep these animals alive, for god's sake, and all they had to do was make the effort to suckle! Frances and the helpers learned to recognise the signs and would come in and take over – even the dog would head for home when he sensed me getting annoyed.

For the most reluctant cases, I called on Tim. He was a shy, quiet chap in his late thirties who loved being around the animals and was particularly keen on the lambs. His only fault was that he used to get very attached to the poorly ones and was sometimes inconsolable when they inevitably died. His great strength however was that he had almost infinite reserves of patience. He had what he called his 'patented method' which involved carefully locking the lamb between his wellies so it couldn't move and then shuffling backwards towards the teat. The first time I saw him do it, I had to turn away as I started to laugh, but it was very effective. Some of our other helpers thought Tim was overly sentimental sometimes, but I came to regard him as a genuine, caring guy who saved the lives of many poorly lambs with his care and attention to detail.

Despite these early frustrations most lambs learned to drink quickly and with the security and warmth of their friends for company, they soon settled down. The smallest ones had their own nursery pen with a heat lamp and a lower level teat so they could reach it easily and they would sleep curled up together for warmth and reassurance in what a friend once called a 'huddle' of lambs. I used to enjoy spending time watching them while on the late lambing shift. The warm, milky smells from the sleeping lambs was rewarding and reassuring after long, energy-sapping days.

All was going fine until others got to hear of what we were doing.

Most farmers look after their stock properly; they care for them, feed them and generally keep them healthy. But they tend to see them as a commodity. I like to think I have empathy with sentient animals that feel pain and fear, but my fellow sheep-keepers would regard me as soft. Looking back, I reckon they thought this was bonanza time – this fool was willing to pay for tiddlers and would take anything he was offered! Hard-bitten business always overrides sentiment in most farmers' eyes, and I think that was the main cause of our problems.

Nobody set out deliberately to damage or exploit us, but we represented a good opportunity to off-load surplus orphan lambs that would otherwise have either died or been a time-consuming nuisance. They saw it as a 'disposal'; we saw it as a business opportunity. We expected the same care and empathy that we would apply to any animal we sold, but sadly they didn't see it that way.

If we'd stuck with our initial supplier who did share our values, we'd have been OK, but we got over-enthusiastic and, flushed with our success with the first few lambs, we agreed to take some from another two farms and set about creating a large-scale tiddler-rearing operation. We spent time building bigger rearing pens and spent money on extra feeding machines. This was just so simple – why bother with all the hassle of lambing ewes when you can just buy the lambs in? 'Why don't more people do it?' we wondered.

The first problem was that our own sheep had started to lamb by then and we were worn out. Weary-eyed from lack of sleep, three a.m. finishes followed by six a.m. starts, meant we were often struggling to keep up with the routine jobs for our own flock. Like all sheep farmers at lambing time, we were stumbling from one crisis to the next, one foot in front of the other, one lambing at a time with little opportunity to grab food or a rest. The last thing we needed was a new tiddler-rearing enterprise to manage.

We coped with the routine tasks like cleaning the feeders and mucking out the pens, but although we tried, we just couldn't give these bought-in youngsters the care they needed, and deserved. This, coupled with the poorer quality lambs from our two new, much less diligent suppliers, was a recipe for disaster. Their disposal rather than sale mentality meant that we were getting lambs that had not had colostrum – the first milk, rich in antibodies which is so vital to any young animal's survival – or that had not been fed properly. And even when they were treated well, no farmer in that frame of mind was going to let us have the bigger, stronger lamb. We always got the poor one, the runt, or the weakest triplet. And as my friend Tim (another Tim) said afterwards, 'You never know why a lamb has been orphaned in the first place. There might be some underlying infection there.' Great advice, after the event! But we were coping, until the big one hit us.

We were just about keeping everything going. Our own lambing was progressing well, the weather was good enough to get lambs and ewes outside quickly and we were keeping on top of the tiddler-rearing until one morning I noticed one or two of them with slightly swollen lips. I could feel the blood draining from my face and my shoulders slumped. Frances looked across the pen at me. 'What on earth is wrong with you?' she enquired.

'This looks like Orf,' I said. 'If it is, we're done for.'

'I'm sure it isn't,' she replied. 'Why do you always assume the worst?'

'Bitter experience,' I thought, but said nothing.

Orf is a nasty viral disease that affects young lambs and causes their lips to swell and blister. It's a bit like cold sores in humans and is very painful and discourages them from suckling. Lambs reared on artificial feeders are particularly at risk as they are all suckling the same teat and the infection can spread frighteningly fast. It's a very damaging disease economically as infected lambs never really recover from the growth check and never 'do' well. And in severe cases they can die.

Over the next few days, the swellings on the lambs' lips grew larger; nasty red pustules formed and started to erupt. First it was just one or two animals, and then four, and then ten, and within a week the whole pen had it. We tried to stop it spreading by fixing solid boards between the pens but we were too late. One after another, whole pens went down with it, spread by all the lambs suckling the same teats. We were surrounded by lambs with multiple blisters on their lips. They stood with their heads down, listless and unwilling to feed, growing thinner and weaker by the hour. They stopped using the feeders as it was too painful for them to suckle and we had to coax them to drink, one at a time, from bottles. Teasing the teat past their swollen lips, using teats with large holes so the milk just ran into their mouths without them having to endure the pain of sucking. We'd have lost most of them if we hadn't embarked on this time-consuming exercise and, to make matters worse, we were endlessly changing gloves and disinfecting everything with bleach in a fruitless attempt to stop it spreading. Plastic gloves, which usually split the moment they got on my hands, were a real pain to use and very uncomfortable. The powder in them turned to mush within seconds from the sweat from our hands, and removing them after each lamb left our hands clammy and cold and made it doubly difficult to put on the next pair. We tried stronger gloves which would withstand the bleach and could be kept on between lambs, but they were too heavy and made the job even harder.

We'd have dispensed with gloves altogether and just washed

our hands with disinfectant between lambs if we hadn't needed to protect ourselves. Humans can get Orf – it's what the vets call a 'Zoonosis' – usually from infected lambs via cuts and abrasions on the hands and it's a very painful condition. One of our student helpers was having none of it and left. She'd contracted Orf while working on another local farm the previous year and was in no mood to suffer the blisters and the pain for a second time.

It was one of the worst attacks of Orf I've ever seen, aided by large numbers of closely housed animals, poor husbandry through lack of time and less than thrifty animals to start with. My initial feelings of despair and depression soon gave way to anger. We had no Orf in our own flock so it must have come in with the bought-in tiddlers. I made a few enquiries and sure enough, one of the two new suppliers had the disease. As do all flock owners with endemic Orf, they vaccinated their lambs at about a week old to protect them but they were sending us lambs that had not been vaccinated. Not only was this unbelievably irresponsible but they hadn't thought to tell us. They had the grace to be a little remorseful and paid for a batch of Orf vaccine for us to use, but by then it was too late. The disease had taken hold in our lambing shed and we just had to keep these incomers away from our own stock as best we could and let it run its course.

The weakest lambs succumbed within a week or two and we were measuring baby bodies by the dustbin full at one stage, but gradually we began to see signs of improvement. The biggest lambs recovered first and with great relief we managed to wean them onto hay and concentrates, so they were at lower risk than those still suckling. We carefully reintroduced the smaller lambs to the milk feeders which released several hours of our time each day for other tasks, but it was at least three months before the infection had gone completely.

By early July the tiddlers were all outside, running around with their contemporaries from our own flock, slightly smaller, with a dull rather pasty appearance and much less boisterous. But at least they were alive and growing. At that stage I began to feel a little more optimistic. Maybe we'd got through the worst. Maybe we'd finally beaten Orf and got a group of saleable lambs for the winter.

But as the autumn rains and cold approached, the tiddlers' less robust constitution started to show. I had the heart breaking job of gathering up sick and cold lambs some mornings and bringing them into the barn to recover. Sometimes I was too late and had to collect the thin creatures stretched out on the dewy grass and sometimes, even with a warm and dry bed, they would still not make it. This was a doubly depressing time as all our hard work and optimism slowly evaporated and it was also expensive as we had to buy increasing amounts of high energy drinks and other remedies to try and keep these lambs alive. Finally in November, we brought them all inside so we could look after them properly and feed them on hay and concentrates. This undoubtedly saved many of them but at significant financial cost. By this stage it was a welfare exercise, trying to keep as many alive as possible with any thought of profit long abandoned.

Frances had long since washed her hands of this project. She'd watched with dismay as these lambs had struggled and gradually faded away as the year progressed and as our book-keeper, she'd seen the huge losses we were stacking up. We'd spent more than £2,000 on milk powder alone, plus all the extra feed later in the year. I'd tried to remain optimistic. We'd made a few mistakes but we were learning from them. And I wheeled out my old standby, 'Nothing worthwhile is easy.' But we were seriously in debt, embarrassed and very chastened. Of the one hundred and five lambs we bought in we lost fifty-seven of them – just over half – but without our interventions and hard work it would have been much worse. In all our years as farmers, this was one of the few times when I really wondered why on earth I put myself through it.

But the real blow was psychological – disastrous decision-making, failure to ask the right questions, and of course all those little lives that didn't make it.

The postscript to this dispiriting tale is rather more heartening. Our farming friends, never short of a pessimistic observation, told us we'd have to vaccinate from then on. 'You're now an Orf infected farm,' said Bert gloomily one day. 'You'll have the chore and expense of vaccinating every year now.' But I was determined to prove him and them wrong.

We kept the Orf lambs inside until all the infected blisters had dropped off and the lamb's lips had healed. This was our attempt to prevent the infection moving out onto the pastures where it would infect our own lambs. We were largely successful with this strategy and were left with clean fields but an infected barn. Harriet told me that although horrendously contagious, Orf wasn't very persistent and was very susceptible to modern anti-viral disinfectants. 'But you'll still have to vaccinate from now on,' was her parting comment, just in case I got too optimistic.

So we cleaned every spec of sheep muck from the barn floor and walls. We brushed and washed and brushed and washed again. I even bought a pressure washer specifically for the job. Once we had it looking as clean as the day it was built, I spent many hours in a very hot and cumbersome full protective suit spraying the whole place with evil-smelling disinfectant. We then left it for a month before allowing any sheep back in. While I was doing this, Frances, in a bid to ensure that 'we were never going to do anything this daft again', threw away or sterilised and sold all the lamb-rearing equipment.

*'Against all the odds, we'd managed to maintain
our farm as Orf-free.'*

Although chastened and embarrassed after this experience, we were much less naive. Not all our suppliers had the same ethics as us and where they saw money to be made, that took precedence over animal welfare and decent business practice. I should have asked more questions, which was lack of experience on my part, and I will forever be uncomfortable about that. Those watching from outside knew the risks we were inadvertently taking but said nothing until it was too late.

If such a debacle can have a positive outcome it was to turn us into more business-wary and worldly wise farmers. And since that day we never had more than one or two isolated cases of Orf each year and in some seasons, none at all. Against all the odds, we'd managed to maintain our farm as Orf-free. A truly unprecedented feat as far as I know and best of all, we were able to prove the nay-sayers wrong!

5. LEARNING THE HARD WAY

IT WAS QUITE SIMPLE REALLY: we needed to sell more lamb. That was the only way we would make the business profitable. And to do that we needed two things – more animals to sell and more outlets through which to sell them. Blindingly obvious but not that easy to achieve.

Breeding and rearing more lambs wasn't really on. We couldn't afford to buy lots of extra ewes and we didn't have the lambing accommodation for them anyway. Our disastrous experience with bought-in baby lambs had been chastening and ensured we would never do that again.

And so, as with so many other things, we eventually came round to doing what everyone else did; that is, we bought in 'store' lambs. These are much older animals that have been weaned and dosed against illness and are generally sold by farmers who don't have sufficient grass to finish (fatten) them.

If you're wondering why on earth I didn't do this at the outset as it seems to be such an easy, obvious solution, I offer two lines of defence. The first is that farmers' market regulations are quite stringent on where the produce is reared, and secondly, store lambs can be expensive – £40 to £60 apiece at the time. The other downside was that I had to buy stores in batches, often twenty or thirty at a time which hit our cashflow quite hard. To quote my long-suffering business partner, wife and book-keeper, 'Our money comes in in tens and goes out in thousands.' So we had to be careful about over-committing ourselves.

We double-checked the regulations with all the markets we attended and found to our relief that we could buy in animals as long as they were on our holding for a minimum of six weeks. But buying store lambs also brought with it a risk of importing disease.

I was working in the butchery one Friday evening getting ready for a busy weekend when Dave, our local auctioneer's dad, appeared at the door. 'Got thirty stores on my trailer for you, Andy, where do you want them?' This wasn't uncommon. Michael, our auctioneer, used to buy for me. I'd tell him how many I wanted and he'd buy them when the price and quality was right. The only problem was, I didn't know when they were coming and on a day like today I didn't have time to go and organise unloading.

'Just put them with ours up the road, Dave, please,' I said, 'I'll sort them out next week when this weekend is over.'

Dave smiled. 'I don't know how you do this! I'm never sure if you're a farmer or a butcher. Still, you're making a lot more cash than we are just farming, so good luck to you.'

'I wonder myself sometimes Dave,' I muttered, 'but it keeps the bank happy.'

Of course I didn't get time for more than a quick check of

'Frances and I spent a miserable afternoon dosing lambs in the snow.'

these new arrivals the following week and it was more than a month later when I noticed one or two very poorly looking ones. I brought them back home as the weather was turning wet and cold and penned them on some nice dry straw in the barn. But it was too late. Harriet the vet was with us a few days later and she noticed them as soon as she walked through the door. 'Are you sure these haven't got scab?' she said with just the tiniest hint of reproach. Sheep scab, to give it its proper title, is caused by a microscopic mite which gets into the fleece and causes sheep to rub themselves on fence posts and walls in an attempt to rid themselves of the itch that it causes. They had got it, of course, brought in from the market, and because I'd brought the ewes into the barn for the winter by then, we had to treat everything. This cost us a fortune but was, thankfully, successful. To avoid further infection we had to treat the remaining lambs outside as well and to make matters worse, by then the weather had really closed in and Frances and I spent a miserable afternoon dosing lambs in the snow.

It was nobody's fault. The scab wasn't evident when the sheep went through the market, but we learned an important lesson, that farming good practice cannot be ignored, however busy you are.

Despite this setback, we found that buying store lambs through Michael worked very well and enabled us to build our retail trade far more quickly than by growing the breeding flock. And, as ever wise after the event, from that day onwards we always quarantined every animal bought onto the farm and routinely treated them all for scab and worms the day they arrived.

FINDING MORE PLACES to sell was more challenging. Our market sales were good but they weren't growing. We could sell more by doing more markets but the takings at each market were static. 'It's not like the old days,' sighed Joe, a purveyor of game. 'I could take five hundred quid without trying here a few years ago. This farmers' market bubble has burst.' He wasn't the best of traders, forever remembered for the comment, 'Blasted maggots – even vac packing won't kill 'em!' But he did a roaring trade in all manner of wildlife, some looking dangerously like road kill to me, and if he was worried then we, probably, should be too.

We'd heard similar tales from other well-established, if less melancholic, stall-holders. The recession had started about the time we began trading but our sales had held up well and we reasoned that loyal, quality-aware customers like ours would cut back on other things before they stopped buying from us. Nevertheless we started to look for other outlets.

Some of our trader friends were attending the food festivals that were springing up in the area and we decided to investigate these for ourselves. I got the details for the Ludlow Festival, the first of its type in the country and arguably one of the best, and sat down with a calculator. It didn't make encouraging reading.

'You do realise, we couldn't get enough lamb in the car

to earn the stall fees for the weekend,' I muttered as Frances peered over my shoulder.

'Oh well, nice idea,' she said with just a trace of satisfaction. She was viewing more days standing behind a stall selling our produce with less enthusiasm than me.

Then by chance, a few months later, I was talking to a friend who was enthusing about the new Hereford Food Festival. 'We had a brilliant day and we had to send someone home at lunchtime to restock. It was easily the best day we've had and the fees are reasonable. It's a new event and they'll no doubt put their prices up now but I'd get in for next year.' So we did. He was right, they did put the prices up but they were still reasonable and they were keen to have us.

We found over the years that festivals are run on an old boys' (or more often old girls') network. If you offer something different, get to know the organisers and ideally get in while they are building up the event then they're more likely to allow you to trade. It most certainly isn't a case of filling in a form and turning up on the day as some believe.

We had a year to wait for Hereford but it proved to be as successful as we'd hoped. We sold out both days and added a sizeable lump to our business turnover that year.

Rather less successful was the Taste of Herefordshire competition that we were encouraged to join as part of the pre-festival build up. I'd spent days preparing the application. I knew how to do these as I'd done them in my other life in ADAS and I put together a huge catalogue of all that we were doing, together with statements of our 'ethos' and customer references. A very polished job, I thought. The Council thought so too and we were shortlisted. I was so excited, as we prepared ourselves for the judges' farm visit.

The chief judge was the affable and friendly Sir Ben Gill, the retired president of the NFU, who now lived locally. He was

Sir Ben Gill and Jane Lewis came to see us.

accompanied by Jane Lewis, Hereford Council's Tourism, Food and Marketing manager. Sadly Sir Ben is no longer with us but has left a huge legacy both nationally from his NFU days and locally from his all-too-brief retirement. 'So where are you going with this business?' was his first question followed by 'And once your ADAS pension kicks in, are you going to carry on or retire?' Both were questions only a person with good business and farming credentials would ask. He could see that the business needed more investment to make it sustainable and that we were going to struggle to get it to a scale where family or others could take it over.

Undeterred though – this was a food festival award after all, not a business competition – we went to the awards ceremony with great expectations.

The Hairy Bikers were in attendance as we all made our way into the very expensive pre-festival dinner and awards ceremony. Our friends had joined us, all paying for their own tickets and hiring dinner jackets for what we had been assured was a black-tie do.

As we gathered in the reception area we noticed that we were the only group in black ties but no matter – this was to be our big night. We had a decent meal which cost £80 per person and then had to listen to a talk by the guests before the presentations started. I was so convinced we were going to win – I even straightened my tie as the placings were read out. We didn't get third place – we were better than that – or second, we were in for the top prize! Which of course we didn't get. One of my guests stood up with her camera to record my victory walk to the rostrum and had to rather sheepishly sit down again and I really couldn't look the others in the eye.

The ride home in the taxi was a very quiet affair. We said our goodnights and went home crushed.

The following day the award winners all had their award plaques on display on their festival stalls. The winners in our category were friends of ours and they graciously explained that it was their 'turn' and that if we kept at it, 'we'd be next'. We let the dust settle but decided to ask for some feedback from the Council, who sponsored the awards. They were very apologetic about the mix-up over black-tie – apparently the person I spoke to (twice, to be absolutely sure) was junior and didn't know what a black-tie event was. Otherwise, they were very accommodating and gave us some useful advice.

But no matter – it was an interesting experience and we landed a large commercial customer as a result. And importantly, we learned not to take friends to award ceremonies unless we were paying and to check and double check that 'black-tie' really is what it says it is!

BUT BACK TO THE FARMING. We had a sizeable operation now and we couldn't just rely on help from family and friends, especially at lambing time. So we decided to start using lambing students from agricultural colleges. The idea was that, as

'Like most serious farmers, we had a suite of lambing pens.'

well as accommodation and meals, we gave them experience and training and they helped us through the most wearying time of the year. It was an interesting and generally rewarding experience, but our most interesting helper wasn't even a student.

Joan was exactly what you would expect a retired ward sister to be: brisk, matter of fact and organised to a fault. She turned up as our stand-in lambing helper one year when our planned student help unfortunately couldn't come

I'd met her six months earlier when she came to see me after one of my WI evening talks about the farm and asked if we would teach her how to lamb. 'It's something I've always wanted to do,' she explained. She seemed very keen and lived only half an hour away, so she wouldn't need to stay with us but quite frankly I had tried to put her off. Lambing is very hard physical work, it requires enormous reserves of physical and emotional resilience, especially when things are going wrong, and it's more about administration, paperwork and tedious routine than clinical excitement. She arrived armed with her book on lambing techniques – I think she was expecting more *Casualty* or *Holby City* than the relentlessness of the lambing shed.

And it is relentless; that's what makes it so wearing – you can't just walk away for a few hours – and you can be sure if you leave a helper on duty while you grab a meal or some sleep, they'll be knocking on the door within ten minutes asking for your help. Big flocks can afford to pay properly qualified staff but we had to make do with unpaid, enthusiastic but amateur help.

Joan found the place unhygienic, and I suppose it was compared to a labour suite in a hospital, but by farming standards we ran a pretty tight ship. We had the old units and sink from our last kitchen with hot and cold water plumbed in. We had a fridge for drugs and milk and were very careful with our cleaning routines for lamb feeding bottles and tubes and we sterilised everything with boiling water and bleach. The place was well-lit and we did our best to keep the floor clean and dry.

Like most serious sheep farmers, we had a suite of lambing pens – more properly known as mothering-on pens because we moved the ewes and new-borns to them after the lambs had arrived. We had twenty-five of these pens which the book says should be enough for a flock of two hundred and fifty ewes but we often got 'bed blockers' where ewes or lambs needed a bit of extra attention or the weather was too bad to turn them outside. So, we often ended up making extra pens from pallets and bits of tin – usually at four o'clock in the morning – to accommodate a rush of new arrivals. The most pens we ever had active was thirty-seven; our student that year chalked the number on the wall behind it – a record never achieved again.

The main job of the students and helpers was to look after the routine things – keeping the hay racks full, checking and cleaning the water buckets in the pens and watching the ewes and new arrivals carefully to make sure everything was OK. Joan was good at this, indeed much better than the younger students who would get distracted by their phones, but she took it upon herself to fetch me for every single real or imagined problem. 'I'm sure this lamb isn't well, I've not seen it feed,' she called.

'It's asleep,' I said for the third time. I know it's better to be safe than sorry but when you've been up all night it can be hard to remember that.

To be fair, she was really interested and helped me with one or two deliveries, but she wasn't really strong enough to pull out the difficult ones. I did what I did with the younger students and let her deliver lambs which would have been born naturally to build her confidence, but she struggled with the muck and crudeness of it all and the fact that ewes don't lie still on a delivery table – they have to be caught and restrained before you can help them. She also had some interesting ideas – including walking round the field with poorly lambs that were about to die just to 'let them see some sunshine' – but she worked hard and did her best while she was with us.

I have to say though, I was pleased and relieved to see Abbie arrive; she'd been before and had asked to come back. She 'had a bit about her' as they say around here and showed a real interest in what we were doing. We've kept in touch with her and she's now a research scientist in human medicine, having found farming science a bit pedestrian; but she was keen, bright and great fun to have around.

Then there was Katy, a smart undergraduate from my old university, Newcastle, who clearly had no intention of ever lambing another ewe after leaving us, but was one of our better students. She took an intelligent interest in everything we were doing and had an indomitable spirit – a great asset during a tough lambing like we had that year. Her only shortcoming was that she was quite diminutive. She made up for that with energy and intellect, but the morning after a snow storm was to be something of a challenge for both of us.

It was just getting light when we clambered into the Land Rover. The wind had blown the snow off the fields and piled it onto the roads, in some cases several feet high. The marshmallow drifts were stunning to look at as they sparkled in the sunshine but our thoughts were on finding a way through the lanes to the sheep on a distant farm and of course on what we might find when we got there.

All was going well until we turned into the lane about a mile away from the fields where the sheep were. It was completely blocked with Land Rover-high drifts.

'That's it then, Katy, I'm afraid we'll have to walk from here,' I said. 'I'll take the buckets and shovel, you just concentrate on getting through the snow and follow in my foot marks.'

'No, I'm here to help,' she insisted. 'I'll take my share of the carrying.' And with a look that said 'Don't argue', she picked up one of the huge twelve-kilo buckets and marched off.

We walked in silence for about half a mile, across a field, avoiding the blocked lane, and branched right, through the woods to the sheep field.

The walking was easier under the trees and we made good progress to the field gate. 'This might not be a great scene,' I warned Katy, trying to prepare her for the worst. I was very worried. I could see us having to dig sheep out of drifts and if the ewes were buried, then the lambs would have died.

But as we approached the gate, we heard the familiar sound of the ewes calling their lambs as they raced towards us. The lambs were running around like children, skipping in the snow, all obviously full of milk. We emptied the buckets in little piles so the ewes could gather round and eat without treading the feed into the snow and stood back to count them.

'They're all here,' called Katy. 'Twenty ewes and forty lambs.' Some folk can count sheep quickly and some can't. I'm in the latter group.

'That's amazing,' I replied. I honestly thought we'd lose some overnight. We walked to the other side of the field to find a line of sheep-shaped bare patches under the hedge. The wind had blown the snow over the hedge where the ewes and lambs had spent the night.

'After the lambing you've had, I can see why you'd expect the worst,' she said, 'but you only put the strong ones out. If I've learned anything from you it's that lambing is about attention to detail. Most farmers would have cleared their pens yesterday to make some room and let them take their chances. You didn't do that and you've not lost any.'

'Yes, you're right I suppose,' I mumbled, not used to praise from the students.

'I reckon we've earned a cup of coffee and a warm-up,' she said. And for once I just smiled.

Probably our most notable student was Ellie. She lived just up the road and was very keen from the start. She could handle all the sheep-related work and because she came regularly, she knew how to keep the records and tag the lambs. She became a very valuable and reliable helper over the years, with the sheep and the butchery and retailing (despite being a vegetarian) and she remains the only one of a succession of wannabe vets who actually made it into the profession.

In the ten years or so since those first fourteen lambs, we'd learned a lot about lambing which we tried to impart to the students. I'm sure the most important lesson was that most ewes know what they are doing at lambing time. We had one heart-breaking case where a ewe contracted listeria (from the mouldy silage mentioned earlier) in late pregnancy and could not stand. Her lamb was born at night; it found its way to her

'Our most notable student was Ellie... the only one of a succession of wannabe vets who actually made it into the profession'

head to be cleaned and then to her udder to drink. It stayed with her for several days and she looked after it as best she could until she died and it was fostered.

Lambs are naturally born in a 'diving' position with two front feet and a nose being the first things to look for at the onset of a normal birth. The most common lambing problems were a front leg backwards or 'hung' lambs when both front legs were back and the head appeared first. In both cases, the lamb had to be pushed back in with lots of lubrication and the legs rearranged so the lamb could deliver naturally. Head first lambs were very nasty if not dealt with quickly as the head would swell and was then very difficult to put back – and there is nothing quite like the feeling of being pulled from your bed at five in the morning to deal with one of these

– you know you're in for an hour or two's shoulder-damaging hard labour.

Lambs coming back legs first usually deliver on their own but there is a risk that they will suffocate as they are born. So the sight of legs and a tail appearing rather than a nose always caused a flurry of activity as we made sure that the lamb was pulled out quickly.

We had our share of real dramas of course. The first time I had a breech presentation (bottom first with back legs forward) was at four in the morning when I was on my own in a freezing cold shed. It's amazing what you can do if there is literally nobody else to help. The lesson I learned that night is that when sheep strain but show no other typical pre-lambing behaviour like pawing the ground or nesting, you need to intervene quickly as the lambs are almost certainly breech.

We always instilled in the students that they did not need to work at night unless they were confident that they could cope, but Abbie did try on one occasion. I woke in the early hours to see the light on in the barn and went to investigate. She was there on her own trying to deliver twins – in tears with frustration but trying to deal with one of the most difficult malpresentations of all. 'You going to let me have a go?' I asked tentatively.

'No, I can do this,' came the terse reply. I had to admire her courage and her doggedness, but there were three lives at stake here so I gently suggested she take a rest.

Both lambs were coming at once and Abbie had been trying to deliver the head of one with the legs of another. It's one of the hardest problem births to deal with and it takes years of practice to learn how to sort this tangle out. The lambs have to be pushed back inside the ewe and then you gently work your fingers along the neck of one lamb until you find it's shoulder and leg and then do the same on the other side. You then have

137

to deliver that lamb while holding the other one back or they both get jammed in the pelvis. It's not like it was on *All Creatures Great and Small*; it's very difficult and requires patience, great care and a lot of endurance.

Our ewes always seemed to find the coldest, draughtiest spot to have their lambs. One Sunday afternoon I was struggling to deliver twins from a very uncooperative ewe in our lean-to open-sided barn with snow blowing around me and freezing water dripping down my neck. I managed to get them out and breathing and got her on her feet so she could feed them. Lambs that are hard to deliver often take time to learn to suckle but it was too cold to risk leaving them so I crawled under the ewe with one lamb at a time to latch them on and ensure they had some life-giving first milk (colostrum). Having rigged up a pen around them with fresh straw and a sheet over the top to keep the snow off, I decided I'd earned a cup of tea in the warm. On returning about twenty minutes later, I discovered that the ewe had laid on and smothered both her new lambs. Such is lambing life, but it's easy to see how it can get you down. Suffice to say, that ewe didn't get the opportunity to do that again. Removing from the flock ewes who are lousy mothers is another hard but very important lesson.

But back to Joan. She'd arrived with her lambing techniques book which I'd politely suggested she didn't really need, but I knew she didn't believe me. One of the last ewes to lamb that year got into trouble; it was a very difficult one where the head was back. These are even harder to deliver than two at once because you have to pull the head forward so it will fit through the pelvis but as soon as you let it go to pull the front legs the head slips back again. I've spent hours trying to correct this and even though I've been shown how to use a so called 'snare' by the vet, I just can't do it. The idea of the snare is to hook it over the lamb's head, below the ears so that you can keep the

'Lambing sheds are strange places.'

head forwards, while you pull on the legs with the other hand. All this has to happen inside the ewe and it's something which requires a lot of practice and skill.

This one was not going to plan. Joan had finished by then but she had turned up with a friend 'to watch'. Her friend was one of our customers who we knew well and I explained to them during one of my increasingly frequent breaks that this wasn't going to end well. After about an hour, I was just about to call the vet when I was aware of a presence behind me. It was Joan with her book. 'Don't you think you should get a bale of straw and lift her back end up onto it like in this picture?' I won't tell you what I said, but we never saw Joan again...

Emily the vet came shortly afterwards and expertly delivered the lamb (with her snare) in about five minutes flat and all was well. I've often thought that vet colleges should teach vets, especially young female ones, to make these jobs look difficult. It's always a bit disheartening to have struggled for an hour trying to deliver lambs and have the vet arrive to complete the job in a few minutes.

'It is, after all, a sheep farmer's harvest.'

LAMBING SHEDS ARE STRANGE places. For eleven months of the year they are quiet – silent even – covered in dust and cobwebs with no hint of the dramas and life and death struggles to come. On good days our shed was a place of hope and excitement and I never tired of seeing new life come into the world and the magic of tiny lambs finding the teat, mostly unaided, for their first feed. On bad days it became a depressing place. We learned quickly not to leave piles of dead lambs around as many farmers do: it's dispiriting for the staff and a constant reminder of lost life and lost income. On these days we just pressed on, one job at a time, one lambing at a time in the knowledge that all these lives were dependent on us and we couldn't give up or throw a tantrum. A good antidote was to walk round the fields of ewes and lambs in the evening – to watch the lambs having their 'lamb races' – always a reassuring sight on even the darkest days.

I don't miss the bad days, the lack of sleep or the relentlessness of lambing, but I do miss the excitement and the real pleasure of things going well. It is, after all, a sheep farmer's harvest, a time when care and attention to detail can affect a business for eighteen months to come. So it's important and at the same time nerve-wracking but immensely satisfying when it works out well.

I'VE TALKED ALREADY about the folly of trying to make our own equipment. Over the years we wasted a lot of time and saved very little money doing this. Top of the bizarre list was the sheep yoke I made with my new welder, a bit over-specified and heavy enough to restrain a small rhino. This was followed closely by our first stock trailer, based on a camping trailer, that was barely big enough to carry three lambs. I used it for several years to take lambs to the abattoir until one day I tried to get four lambs in and we burst a tyre on the way....

It took us a long time to learn that buying decent kit saved us time and money in the long run. Our best buy was a dagging crate that cost £500 but it was money very well spent. The ewes walked up into it, were restrained by the neck and the whole contraption would swivel so that the sheep's rear end was at my waist height. This made it possible for me to crutch two hundred ewes on my own in a day and treat them or inject them at the same time.

The other huge lesson we learned was that introducing electronic identification into our flock was far from straight-forward. I've always been a bit of a nerd when it comes to electronics and computing. I was the farm computing special-ist in ADAS from the earliest days of PCs back in the early eighties and so I was probably keener than most to try to use the technology to manage our flock. And we were also still on the constant lookout for things that would save time and make us more efficient.

Frances was sceptical as usual, more expense with doubtful

benefits was her assessment but the students were more receptive. 'Electronic identification has three components,' I explained. 'There's a tag which is inserted in the animal's ear, using pliers, just like an earring. That tag contains a code which can be read by a reader that looks a bit like a walkie-talkie, and the same gadget can then be used to record information about that animal.'

'Hey that's great,' enthused Abbie. 'So we'll have all the information at our fingertips right here in the barn.'

'Not quite,' I cautioned. 'The information has to be uploaded onto the PC in the office as the reader we use in the barn only holds a very limited amount of information. It's more about identifying the animal and recording data.'

'So what we really need is a laptop we can use over here then?'

'Well, yes, but they're far too expensive. Let's just do the recording bit to start with and see how it goes.'

It was Abbie's turn to look sceptical but we carried on.

Electronic tags come in sets: an electronic one which is always yellow and goes in the left ear, and a visual one which can be any colour and goes in the other ear. We used the different colours for ewe lambs we wanted to keep that were born in different years so we could identify them easily.

We had our first problem here. Tags for lambs have to be small or they droop in the ear, and they have a small peg which is less painful for baby lambs. These lamb-size tags with their small pegs, however, pull out of older sheep's ears and are often lost. We therefore needed two sizes of tag and for the ewe lambs that we kept, we had to change their tags when they joined the breeding flock.

Inserting tags into sheep's ears was the next challenge. Ewes' ears are very tough and we needed a very good tag plier to get them in. Lambs' ears are much softer, but they wriggle

142

'Our first tag reader was a very basic affair.'

and it's very easy to break the peg as it goes in. It's always the second tag that breaks – and as they have to be in matched pairs, this then means you have to remove the first one so you can replace it with another pair.

Tagging baby lambs is not without its critics. Some of my farming friends were against it on welfare grounds, and even the vets were sceptical because of the risk of infection. For this reason, one year we didn't tag the lambs straight after birth but bought them back inside with their mothers a couple of weeks later to do it. This was an absolute pantomime. We had marked the ewes and their lambs with a number so we could match up lost lambs out in the field, but the numbers faded quickly and after a few days were very hard to distinguish. We spent most of a morning matching lambs to mums and tagging them – and still had several 'spare' lambs left over at the end. Another lesson learned the hard way.

Our first tag reader was a very basic affair but it did allow us to record codes against each animal and these codes were completely flexible. We got on quite well with it but unfortunately the supplier went bust so that experiment was short-lived.

SHEEP CAN BE EXASPERATING creatures at times but I've always believed that they're not as daft as some would have you believe. But what I really wasn't ready for, was sheep that could simply disappear.

'Thanks guys,' I called as I peered into the empty trailer. I was at the abattoir and the men had unloaded my sheep for me while I completed the interminable paperwork. I'd never understood why animals needed a licence to travel; it's not as if they could drive. Even humans don't have to be identified individually when travelling, but animals do.

The lads always laughed at me with my red clipboard with all the forms, but Paul the stockman had a different view. 'Most of the folk who come in here have those forms stuffed in their pockets, covered in muck or worse. And most of 'em can't write properly anyhow. You stick with your clipboard!'

I had a good relationship with the abattoir staff. You need to be a certain type of character to work in a place like that, but they were good at their jobs and treated the animals with care. I saw them every week, on a Monday to take the animals in and on a Friday to collect them, and we had a 'robust' if sometimes ribald exchange most weeks about the quality of my stock. I'd even got them to identify the individual carcases for me when I needed to get some feedback on a specific animal.

'You don't make it easy for us, do you,' exclaimed Rob, one of the slaughter-men one morning.'

'I don't see why it's so hard,' I replied. 'Williams's could do it. They have a little metal tag on the carcass so each one is identified. It's all computerised.'

That was the wrong thing to say. 'Well you'd better go back to them then and stop messing us around!'

'You know I can't do that – they've gone bust.'

But they were only arguing to make a point and they did it for me without too much grumbling. They had a weight label tied to one leg of each carcass and they wrote the animal ID on it with a felt-tip pen!

But this morning there was no joking and joshing.

'What are you on about Andy, we've not unloaded anything.'

'But you must have, the trailer's empty.'

'We haven't.' I could see them getting defensive.

'Please go and check,' I asked carefully. They thought I was accusing them of taking the animals off for their own use.

They went back inside and a few moments later the foreman, Alf, came out smiling.

'You daft bugger, you must have arrived with an empty trailer.'

Now I was getting cross. 'I know I didn't, I distinctly remember loading them, they were the first shorn lambs I've ever drawn (selected for sale) and I thought how good they looked.'

Alf's smile faded. 'Honestly Andy, it happens quite regularly. You go home and I guarantee they'll still be in the pen. It's usually the phone ringing or a visitor who distracts you when you're loading and because you do it every week, you're on autopilot. So you finish the call, shut the trailer and drive off. Honestly, it happens all the time!'

I was far from convinced but I'd got no choice. I told Paul that I'd bring them in later so he'd better keep the paperwork to one side for now and I set off for home.

An hour later, I rattled into our yard, parked up and made my way into the barn. Sure enough, just as I thought, the pens were empty and there were no sheep waiting to be picked up.

I walked back to the house feeling puzzled. 'You're either going senile, which is quite likely, or they've stolen our lambs,' said Frances, helpfully.

'There is another option,' I said without much conviction. 'Could they have got out en route?'

'And how would they do that? Even if they did get out, someone would have seen them.'

'Maybe the little side door came open and they got out that way. It did open by itself one day, if you remember?'

'But you said you'd fixed that!' I was losing the argument. 'Maybe someone opened the trailer side door in the abattoir yard and they're running around the village.'

'Well I suppose that's possible,' I conceded. 'I'll call them and ask them to check.'

I got Doug, the boss, this time. He was laughing too. 'Andy they're not here!'

'But they're not here either,' I said crossly. 'I know it's only three lambs but it's our week's meat for the markets and I'm stuck now.'

'I'll sell you three lambs, at wholesale rate as you're a good customer,' said Doug, ever the businessman, conveniently failing to remember that on the very rare occasions that I bought in extra meat I always paid him the wholesale rate anyway. We agreed a deal and I went back to pondering what had happened.

'Did you stop anywhere on the way?' asked Frances. 'If they'd got out on the main road, someone would have hooted you, so it must have been in the lanes around here.'

'Well I did slow almost to a stop up by John's place. He was moving sheep across the road. But he'd finished by the time I got close. If ours had got out there, they could have got mixed up. He's been shearing too, but how on earth did they get out of a closed trailer?'

My fame spread quickly. Within days, I was getting teased by other market traders as 'the sheep farmer who can't count.' My farming neighbours also got to hear about it and were similarly amused. Bert saw me in the lane. 'I hear you've been driv-

'You know those old trailers are supposed to have a bar across the back, above the tailgate?'

ing around with empty trailers! You won't make much money doing that!'

'I know,' I smiled, 'but I'm sure I loaded them and they weren't there when I got to Doug's.'

'You know those old trailers are supposed to have a bar across the back, above the tailgate? The modern ones, like mine,' Bert was always one to rub it in, 'have a fixed bar but the old ones were removable.'

'Ah,' I said, 'I've often wondered what those two latches are for.' The light was beginning to dawn. 'Do you really think they'd get out over the top?'

'You were shearing at the weekend, weren't you? They're much more agile without their wool on. They might have scrambled out.'

We walked up into our yard to inspect the trailer and sure enough, there were a few traces of muck and straw right on the top of the tailgate.

'Reckon that's your answer,' smiled Bert.

I thanked him and went back to the office to check our insurance. I seemed to remember some clause about 'stock in transit'.

Several days later I got a call from the insurance company's loss adjuster. He was a local farmer who I knew of vaguely.

'What's this about escaping sheep,' he laughed down the phone.

'I know it sounds ridiculous,' I said, 'but it's genuine, honestly!'

I explained what had happened and he listened carefully. I waited for him to burst out laughing but he didn't. 'Sounds plausible to me,' he said, 'and the fact is you don't have the sheep anymore and you were insured against loss. What were they worth?'

We never did find out what happened to those lambs.

WE HEARD ON THE LOCAL grapevine that Bert was planning to retire and was hoping to let his farm land and buildings. With his fields just across the lane from ours, it was an ideal next step but it would be a big commitment and would need more finance. Our bank, which I'd been with since I left school, was spectacularly unhelpful. I prepared budgets and cashflows, just like in my ADAS days, but we were too small and they weren't interested. An old ADAS contact did at least get us a visit from the local branch staff whose only contribution to the discussion that morning was, 'Do you have to get up early?'

After this unhappy experience, we had no choice but to change banks. We talked to a competitor who sent their agricultural specialist out to see us – a guy called Dave whose first question was, 'Let's have a look at the sheep.' It was a huge breath of fresh air. We then, over numerous cups of coffee, went through the figures and he seemed impressed.

The next day, I was on my tractor making silage when he called. 'I like what you're doing and I'll support you with the lot on overdraft' was his opening offer.

'Great,' I said, trying to balance a mobile phone on my shoulder while bouncing across the field. I reflected afterwards that this was one of the more important calls I'd taken in my business life and maybe I should have stopped the tractor to talk to him! Dave proved to be a really helpful and friendly chap and he became our guide and confidant for the rest of our farming life.

With the funding fixed, I went to see Bert. He knew us well enough to know that we would look after his land and that we could pay our way and he readily agreed that we'd make good tenants. Shortly after this discussion, he asked us if we'd be happy to use a land agent to handle the deal. We agreed to this as we thought it would ensure that we had a proper contractual arrangement. The agent then persuaded Bert to hold an informal tender which meant that we would have to submit a sealed bid against other would-be tenants – a nuisance but Bert still insisted that he wanted a local person whom he knew.

Our farming friend Cecil came to see us and suggested that we could submit a joint bid for the whole farm. We warmly welcomed this offer as it greatly strengthened our bid and gave us the opportunity to integrate our livestock farming within an arable rotation.

Frances and I walked the farm one afternoon, planning how we would move the stock around and where we could gather and handle the lambs. I could just see us out in those fields. It was a very exciting time and all the while Bert was reassuring us that he wanted it to go to someone who would look after his farm and not wreck it with overstocking, and that he wouldn't necessarily take the highest bid.

We had the farm rent professionally assessed and then increased our bid by another £30 per acre as it was next door – indeed we did everything we possibly could to be fair and to secure our first proper farm. I even agreed to buy fifty extra ewes from Bert to stock our increased acreage and of course to oil the wheels of the deal a little.

I told bank manager Dave of Bert's comments about wanting a good, reliable tenant and not necessarily taking the highest bid and his response was a quiet smile. He knew what a big deal this was for us.

Tender day arrived. Cecil and I had completed the form the night before and I drove up to the agent's offices with a sense that this was the next step. Michael the agent welcomed me with, 'I think you have an important document for me,' which all added to a feeling of confidence that had been growing all week. Just a few hours and we'd be serious farmers, on a proper farm, accepted by our peers and welcomed by our neighbours.

AND, OF COURSE, DAVE WAS right – we didn't get the farm. We were outbid by a sheep farmer and dealer who offered a huge sum for it and then proceeded to stuff it full of lambs and cull ewes. Lorry load after lorry load passed through every week, bought in from local markets and then shipped off to the abattoirs in Birmingham. That lorry held more animals than our entire annual lamb crop; at busy times they were shifting more than a thousand lambs a week.

The new tenants did what they did very effectively and I had to admire their energy and persistence, but it was a hard lesson learned. Money usually talks.

I was told the news by Cecil's son Ryan who was better-connected in the farming community than me. He knocked on the door late one evening with the news. I walked into the kitchen to tell Frances and slumped down in my chair, completely deflated. Frances patted my shoulder and went to bed, leaving me staring into space.

It was a devastating blow and at the time it seemed terminal. We had lots of commiseration from our neighbours and friends. The farmers among them were furious; some had even refrained from bidding because they thought we should have it.

Bert didn't trouble to tell me officially until some days later.

The postscript to this is that I wasn't quite so foolish as to put all our eggs in this one basket. I had spoken to two other local landowners from whom Bert had rented grass that he was now giving up. Between them, they had almost the same acreage of grazing as was on Bert's farm, and although it wasn't as close by, it would have made a good second option.

Within a few days of our failed tender, one owner decided that he didn't want us to have the grass because his vegetarian wife didn't want meat animals grazing it any longer, and the other decided to plough his grassland up to grow corn. So, in the space of five days, we lost our bid for the farm and both alternative fall-back options. It hadn't been a great week.

It was heart-breaking, a demoralizing blow to us. And it took us a while to recover. Yet again, our assumption that those around us were supportive was proved wrong – or maybe we were just naïve. Bert had every right to let his land to whoever he wished and maybe we'd taken too much for granted. But he had implied that he wanted us to have it and had even set up the original conversation with Cecil for the joint bid.

My initial reaction was to tell him I didn't want his fifty ewes but I decided that was churlish. 'I'll need your trailer to move them,' I said crossly. 'There's too many to fit in mine.'

'OK Andrew,' he muttered (he never did call me Andy like everyone else), 'it's in the yard but I've not cleaned it yet.'

'No worries,' I said, spotting an opportunity. I moved the ewes and then spent an hour pressure-washing his trailer. I returned it, sparkling clean, and he did at least have the grace to look sheepish. I was so angry but I reasoned that I'd got to keep living and working next door to Bert. 'We're bitterly disappointed over this,' I said, trying to sound positive, 'but if this all falls through for you, do come and talk to us.'

The new tenants did complete the deal and we didn't get another chance. But I wasn't going to be put off that easily. I'd find another way.

And Frances? Endlessly supportive as usual, but on reflection, I think she was secretly quite relieved.

6. FINDING OUR OWN WAY

TAKING ON BERT'S LAND was to have been our 'graduation' from hobby farmers to farming for real. It was tempting to walk away at that stage, but I wasn't going to give up that easily.

There had to be other ways to grow our business; our Achilles heel was still our numerous small blocks of land and maybe we should focus on fixing that? The paperwork was also getting onerous and perhaps we could have another go at electronic ID which would certainly save us some time. And maybe it was time to try some of the diversification options that we'd been mulling over for years.

We set up a meeting with bank manager Dave and chatted around some ideas. He agreed that we should look for bigger blocks of land closer to home and use some of the overdraft to buy a good stock trailer and a four-wheel-drive vehicle to pull it. This was a long-overdue decision and it changed our lives dramatically. It meant we could move large batches of ewes and lambs around with comparative ease. My little two hundred pound trailer was consigned to the nettles where it languished for several years until I found someone who was just starting out – a bit like us some ten years earlier. I suppose I should have warned him it wasn't big enough but he seemed keen to have it and persuaded me to part with it for just £50!

We shamelessly exploited the sympathy we'd received from our farming neighbours to get onto the grass rental grapevine. There was a lot of support out there for us and we were keen to move on quickly. One of my better decisions had been not to tell our current landlords that we were bidding for Bert's land, although if we'd got it, we'd have given up all the small tenancies. I felt a bit guilty about that but when I told them after the event, they were, to a man (and woman) very sympathetic.

Our landlord at Hennor, who was especially annoyed on our behalf, decided to help us improve one of her fields so we could make better silage. This turned into quite a project, with my contractor spreading muck and ploughing during the winter and the landlord's neighbour cultivating it the following spring. I then did the rest of the work myself and toiled away down there one late March day in the sunshine, harrowing, rolling, fertilising and seeding, trundling back and forth to home to collect fertiliser and lime. With only one tractor, it took forever, but I got it all done and finished rolling in the seed as it grew dark. As I closed the gate in the gathering gloom it was an absolute picture, beautifully smooth, neatly rolled and ready to grow. And as is the fickle nature of March weather, two days later it was under several inches of snow!

But it grew away well and we had our first high quality crop of silage from it later that year. It was amazing stuff – 'Almost good enough to eat yourself,' as Harriet described it. Ironically,

'I toiled away in the sunshine, harrowing, rolling,
fertilising and seeding.'

it was so good that we had to restrict the ewes' access to it for a couple of weeks prior to lambing as they were gorging on it and I was scared we'd have more prolapse problems. But we didn't, and we had a cold spring that year so this 'rocket fuel' silage, as the feed rep called it, came in very handy until the grass started to grow.

This reseeded grass was a small step in the right direction and it helped us a lot but we still needed more land to graze.

A few months later I had a call from Cecil. 'Come down and have a chat sometime,' he said. 'We'd like to help you out with some grazing if you're still looking for it?'

'Of course,' I responded, trying to sound businesslike and not over-enthusiastic. I had no idea how much land they had or what they'd want in rent. These guys were experienced farmers and they'd know what the land was worth to me.

I went down to see him and Ryan the following day and was delighted to find that they were offering me all their grazing grass. They were arable farmers really but had a number of grass fields unsuitable for cropping that they'd been using to make hay for sale to horsey folk. But this was a lot of work and reliant on the weather so they'd decided to concentrate on their arable work and let someone else manage the grass. In total there were about thirty acres, all connected by tracks or lanes so we could move the sheep around easily without having to use the trailer for every move.

'This is great,' I muttered, trying to play my hand very carefully, 'But I really can't run to more than £80 an acre.'

Cecil laughed. 'Ah, well we were planning to charge you sixty! The sheep muck will do the ground good and we'll need you off in the winter. Don't over-graze it and no hay or silage.'

We did the deal and took on our first big block of manageable land. Cecil helped me put up a handling system so we could gather and treat the animals without bringing them home and we set about expanding our flock to make use of this new ground. At last someone seemed to be on our side.

We got to know Cecil and Ryan well over the years. They were interested in what we were doing, and knowledgeable. They'd occasionally help us move animals around and Cecil would always stop for a chat if we met in the yard. They became – and still are – good friends and I like to think we contributed to their lives in the way they did to ours. We spent many happy hours sitting around their kitchen table, putting the world to rights.

One particular benefit of working with farmers like Cecil was their contacts. He introduced us to a new contractor who could bale straw for us and make what he called 'small big bales'. For those in the know, or who even care, they were half-sized Hesstons!

We'd been baling small bales of straw, what farmers call conventional bales, or more colloquially, 'idiot bricks', be-

'For those in the know, or who even care, they were half-sized Hesstons.'

hind Cecil's combine for years but had always struggled with timing and the weather. Like all arable farmers, Cecil wanted the straw baled and off the field immediately the corn was cut so he could get on with planting the next crop, but it was almost impossible to predict when this would happen. And with a retail business to run, I was often elsewhere at a market or festival when the call came and I'd have to reorganise my day and dash home to get the straw baled.

These 'small big bales' were about the size of a fridge and could, just about, be moved on a sack barrow, so they were a good compromise between efficiency and speed at baling time and ease of use in the barn in the winter. Using a contractor also meant that I didn't need to be there myself but we were still at the mercy of the weather.

'I'm going to get this straw baling properly organised, this year,' I announced. Frances frowned with that knowing look.

'What makes you think it'll be any different? It's always a mad rush with you running in ever-decreasing circles.'

'Well, I know the contractor now, I can get him booked up in advance and Cecil knows we're busy. I'll sit down with him and get it planned properly.'

Frances's face said, 'Good luck with that.' But I set off to find Cecil and put a call in to the contractor.

A couple of weeks later, I was sitting in my tractor on the field headland or field edge. I'd lined up my two trailers and one of Cecil's so I could load and shift the bales as fast as they were baled. There were menacing black clouds on the other side of the valley but I was feeling confident. Cecil was 'opening up' by going round the headland with his combine. I didn't usually have headland straw as it could be wet and Cecil liked to chop it and plough it in. The contractor would be there in a few minutes, in plenty of time for when Cecil started to cut the main part of the field.

Combines have their own smell: a mixture of diesel, dust and warm grain, evocative of those school time harvest days, decades earlier. The clouds were getting blacker as the combine started on the long runs from side to side. I couldn't see the machine for parts of each journey as it was down in the valley but I could see the dust rising above it. Cecil was leaving lovely straight rows of crackly dry golden straw for the baler, exactly what we'd need to keep the sheep clean and warm in the winter.

I was feeling very pleased with myself, we were properly organised for once, I could get the straw loaded and either sheeted or home in the next hour or two, well before the rain was forecast. Then the phone rang. It was the contractor. He'd had a breakdown and wasn't coming. 'Sorry Andy,' he said – we were on first name terms but much good that'd done me. 'Let me know when you've more ready to bale and I'll come and do it straight away.'

'OK,' I replied wearily. I trudged across the field to where

Cecil was working and clambered up the combine steps to give him the news. He smiled grimly. 'That's a shame. You'll have to move those trailers again now so we can chop the straw as soon as this rain clears. There'll be one more field for you to have a go at later in the week.'

I walked back to my tractor and climbed into the cab to take the trailers home. As I turned the key, the first rain drops splashed on the dust on the windscreen. Within seconds I looked on helplessly as the rain swept across the valley, and watched my bright golden straw become a sodden mass. So the weather had scuppered our plans anyway, despite the breakdown.

We did get some straw baled eventually and, with a large purpose-built trailer borrowed from another neighbour, we got it all shifted home in double-quick time. 'We were better organised this year,' I explained to Frances, 'but we were unlucky with the weather.' She just smiled.

WE WERE GROWING in confidence as our standing in the farming community increased. I felt involved and accepted and at village social events I was part of the group who stood and talked farming – a very exclusive set indeed.

A lingering problem was that I still had this battered old tractor while my contemporaries were all driving smart modern kit. 'At least yours is paid for,' was Frances's view, but I wanted something a bit better. Things had come to a head some years earlier when we'd made some really good silage. Nice and leafy and green, the sort the ewes would fight over in the winter. I drove into the field of newly wrapped bales as Phil the (then) contractor was packing up to move to his next job. I was met with that slightly weird smell of drying grass and plastic, not very bucolic but a sign of a job well done. It was a Friday afternoon and I was looking forward to carting bales back to our yard at a leisurely pace.

'This ain't bad stuff Andy,' laughed Phil as he drove away – praise indeed from him. 'Just be careful and don't split the plastic or you'll ruin it.' I drove up to the first bale with my bale 'squeezer' on the tractor and carefully clamped it on the underside of the bale. I pulled the lift lever and nothing happened. I gave her some revs and tried again. Still nothing. With a sinking feeling, I tried another bale in the forlorn hope that the first was one from the outside row and was greener and therefore heavier than the rest. But no, they were all too heavy. My old tractor just wouldn't lift them. Now I was in real trouble. I had a field of forty bales to get back to the farm and I couldn't get them onto the trailer or off again at the other end. I drove home cursing and trying to figure out what to do. By the time I arrived in our yard, I'd decided to eat humble pie and go and see Bert. This was before the problems over the letting of the farm. I explained my predicament and he smiled. 'Yes, you can borrow my big tractor over the weekend, but I need it back on Monday. Do you want my big trailer as well?'

'Thanks Bert, I really appreciate this,' I smiled, pleased to have helpful and supportive neighbours.

I dashed back to the field with his tractor and trailer and started to load the bales. They really were incredibly heavy and I even got his tractor up on two wheels at one point – something he'll never know now as sadly he's no longer with us. I managed to get them home and stacked over the next couple of days but it was a real struggle. And it wasn't just a one-off problem. Now, as a result of this episode, I had to wait in the field with my tractor and loader, every time the contractor started to bale silage. I then had to check that I could lift the first one or two – and get him to reduce the weight if I couldn't – before I allowed him to proceed with the rest of the field. Yet another inefficient and time-wasting exercise that we needed to fix.

'If you buy a good one, you'll be able to sell it for almost

'It did all that I asked of it – and played Radio 4 while doing it.'

what you paid for it,' explained the salesman from the local
Massey Ferguson dealer. 'And I'll give you a good deal on your
old one as they're still in demand on the smaller farms where
they like levers and pedals rather than switches and dials.' And
so began the saga of the expensive replacement tractor. It was a
bridge-crossing decision in that I couldn't undo it but it wasn't
really anyone's fault, it was just bad luck all round.

If I'm honest, it was an indulgence. I was getting fed up
with my old machine: it was uncomfortable and noisy, cold in
winter as well as not being strong enough for the jobs I needed
it to do. It was also starting to cost money in repairs – ironic as
it turned out, but more of that later. I justified it, to myself at
any rate, as providing me with a safe and comfortable machine
in my last few working years. I could use a small lump sum
that was part of my pension but wasn't worth investing. As the
dealer said, I had literally nothing to lose. And as reassurance,
the trade-in deal he offered me on the old tractor was the same
price as I'd paid for it, ten years earlier.

'With only a slight twinge of regret, I watched my old faithful machine go off on the lorry.'

So, with great excitement I went off to the dealer's yard one Saturday morning and did the deal. A '57 plate Massey Fergusson 5435 with a new loader and lots of novel features like glass in all the windows, doors that fitted and a radio that not only worked but could be heard over the engine noise. This sparkly, shiny machine was duly delivered one early summer morning and with only a slight twinge of regret, I watched my old faithful machine go off on the lorry to its new home in Wales.

Within an hour I'd realised they'd fitted the wrong loader – a much smaller, lower spec one than I'd paid for. And within two days the machine had stopped in a field corner and couldn't be revived even by the fitter who came to minister to it with his laptop. Both problems were resolved quickly and efficiently and I had a much higher spec machine as a 'courtesy tractor' while mine was being fixed which was great fun and even more self-indulgent.

Things went very well after that and it did all that I asked of it – and played Radio 4 while it did it – until one day I was moving some silage for my friend Kevin when he noticed a knocking noise. I had noticed it too but thought it was fine. After all, it had come from a main dealer and had been fully checked over in their workshops. As we'd both noticed it, I thought I'd better get it checked out and on investigation it proved to be a problem that needed major surgery – just out of warranty of course. To cut a very long and frustrating story short, the repairs cost £5,000 which was about a quarter of the full value of the tractor and ten times what I'd spent on repairs to the old one in the whole decade that I'd had it!

Having spent this much on repairs, I couldn't afford to change it and had to accept the dealer's word that it was now 'as good as new.' To be fair, I think it really was. It did all our field work and hauled and stacked silage and I used it daily through the winter on feeding and yard work.

The postscript to the tractor saga is the status it conferred not just on me but on our business, with the locals. 'You've obviously made it now and are here to stay,' said one. 'We'll do business with you now,' said another as he saw me driving it down the lane. So a machine bought with my pension fund, which would have otherwise bankrupted the farm, somehow gave me financial credibility and indirectly led to our next expansion step.

KEN WAS ANOTHER NEIGHBOUR I'd got to know well over the years. Always with a smile on his face, he'd been a farmer and ground works contractor. He ran a catering business, did a bit of machinery trading and had even been landlord of the local pub for a while. I liked Ken and spent many happy hours chatting with him in our local over a beer or two. When he'd sold his farm, he'd kept about thirty acres of grassland which he'd let to our neighbour Bert, but when Bert gave it up, Ken had wanted to plough it and grow corn. This was to avoid the restrictions that would arise if Defra (the agriculture ministry) classified it as 'permanent pasture'.

These regulations were designed to prevent damage to permanent grassland, a unique and rare habitat whose value we are only now beginning to understand. But the definition of 'permanent', dreamed up by those who should know better, was five years or older, which meant that a crop of sown grassland (called a ley), left down for more than five years could then no longer be returned to cultivation. So not unreasonably, farmers like Ken were keen to avoid stumbling into this restriction by accident.

A couple of years later, I noticed that Ken had put his fields back down to grass and I sensed an opportunity. He'd let it to Tim, a local dairy farmer, for silage but an overheard conversation in the pub some weeks later implied that he would be giving it up. I had a quiet word with Tim, who I knew well, to make sure he didn't want it anymore and then approached Ken. He'd already been approached by yet another neighbour but he agreed to let me have the two fields nearest to us (about half of it) and the other neighbour would have the rest. I agreed immediately to take it on that autumn with a view to making silage up there in the spring.

As silage time drew near, Ken saw me in the yard and offered me the whole thirty acres, for the price I'd agreed for half of it. Apparently the other neighbour had decided not to follow through and this gave us the opportunity we needed to really move up a gear. Suddenly we had a thirty-acre block of good quality silage grass at a very decent price. Our contractor Kevin was delighted to be able to work on proper 'farm-sized' fields rather than our tiddly meadows, and he duly produced three hundred and two wrapped bales for us one warm sunny day in early June. I laboriously carted them to the field edges and netted them to keep the birds off. I finished at ten-thirty

*'Suddenly we had a thirty-acre block of good
quality silage grass at a very decent price.'*

that evening and drove home in the dark feeling very pleased
with myself. This was proper farming and I was doing a good
job with decent equipment.

Unbeknown to me, Ken's mum had been watching me on
my new tractor moving all these bales into rows, and she'd told
him that he'd better offer me the ground the following year.

Ken did indeed take me on as his tenant. 'Mum thought you
did a good job and now you've got some decent kit, I know you
can look after it,' was his verdict.

I was now left with three hundred bales of silage and the
problem of getting them all home. I sold some in the field for
others to move but still ended up with a mountain of two hundred bales at home after several long days hauling and stacking. This was about double what we needed and I wondered if
it was worth trying to sell some. 'You hang on to it,' was Ken's
advice. 'You can always sell it later – you don't know how hard
the winter's going to be!' Wise words as it turned out – we

struggled with a long winter, used more than planned and sold the balance in the spring.

WE WERE UP TO ABOUT two hundred ewes by this stage which was proving to be a challenge. We didn't really have enough room to lamb them all inside, and it was always a struggle if the weather turned wet and we couldn't get them outside as soon as they'd lambed.

I changed the layouts in the barns many times over the years, trying to make best use of the space and to provide adequate feeding arrangements for our growing flock. The original plan was to have six twenty-by-twelve-foot pens, three each side of a central twelve-foot-wide passage, each holding about twenty ewes. These pens had a feed fence that the sheep could put their heads through to feed along the front and were separated down each side by walk-through troughs – a trough built into a fence that you can walk along to put out feed – to maximise feeding space. This worked well while we fed hay and concentrates, but it became a chore once we had to unroll and fork silage and, of course, the central passage was a hopeless waste of covered space. Trying, unsuccessfully, to learn from this mistake, we had the opposite problem in the new lean-to where I decided to put in a feed fence with a small feeding passage to save space, and ended up having to barrow silage along this passageway twice a day.

After several re-thinks and following the usual theme, we eventually settled on using ring feeders for silage, just as everybody else does. But unlike others, we were very strict about having enough feeding space for all the ewes to feed at once. This is a significant welfare issue in that the bigger ewes get to eat all the good stuff and the smaller ones get left with the rest. As Harriet put it, the fat ones get fatter and the thin ones get thinner. This meant we might have up to four rings in a large

'These pens had a feed fence that the sheep could put their heads through to feed.'

pen and although I was initially worried that they wouldn't eat it quickly enough before it spoiled, it proved to be an easy and very satisfactory system. The only issue was that it needed me or a reasonably skilled tractor driver to drop the silage bales into the rings in such a confined space, and we had to devise ways of penning the ewes back so we could get the bales in without the inevitable 'scrum.'

We also fed concentrates as cobs or rolls on the floor, which meant we could do away with the walk-through troughs. Rolls are about the size of large thimbles and, unlike the much smaller pellets, don't get lost in the straw in the pens. Harriet wasn't very happy about this for hygiene reasons but it worked as long as the litter (straw bedding) was kept clean; its great advantage was that the feed could be spread widely to make sure even the shy ewes got a chance to feed.

With increasing ewe numbers, we needed to buy concentrate feed in bulk because it was substantially cheaper per tonne. I built a couple of storage bins in the corner of the barn with loading pipes so that lorries could blow the feed straight in. This saved us hours of time and many hundreds of pounds a year. As ever, they were home-built as we did not have the funds available to buy proper outdoor, metal bins.

The one thing we got wrong initially was the water supply for the pens. We used small (three foot) water troughs and I always installed them temporarily each season using miles of hose pipe and lots of leaking connectors. This gave us absolute flexibility of pen layout but meant we had to remember to turn on the taps to fill them – and more importantly turn them off again! We only had two serious floods, the worst when I'd been in a hurry to catch a train to London. Returning nearly twelve hours later, I found the whole barn awash and it took several hours to clear up and re-bed the pens.

The other problem was that the water troughs needed cleaning regularly, which was another chore. Harriet used to say that sheep should have access to water clean enough that 'you'd drink it yourself.' And in the end we invested in wall mounted cast iron sheep drinkers, properly plumbed in so that the ewes always had access to clean water. Retrofitting these drinkers meant digging up the barn floor to install the pipework, but it was time and money well spent – and kept the vet happy as well.

But whatever we did, we were always struggling for space.

'We should put up a polytunnel like Paul, was my opening shot. Paul was our ram breeder.

'And how much would that cost?' Frances was seeing pounds with several zeros going out again.

'Oh about three grand,' I said airily, 'but it would be so much easier. We could set it up so we could feed around the outside and not have to move the silage bales into the pens like we do in the barn. We could get more ewes to make better use of it...' I could see she wasn't convinced.

'But it won't be three grand, will it. There'll be water and electric to install and penning and you'll want a track round it if you'd going to feed outside. Is it really worth it when we can buy in store lambs to top up our numbers for the retail trade?'

I had to concede she was right. Looking back, it was a pivotal point in our business journey. And it was a business now rather than a farming operation. I was trying to grow the farming because that was what I enjoyed and I wanted to be seen to be farming 'properly' on a realistic scale. But it was the meat retailing that was making the money and we had no hope of growing the farming to a scale where it could be profitable in its own right.

As an exercise for a talk I gave to the local National Sheep Association group, I decided to cost the farming and retailing separately, charging the retailing business the market value for the lambs as if they were sold in the local market. And it was no real surprise to find that the business would make more money if we stopped the farming and just bought in the lamb straight from the abattoir. That was a bit disheartening but we were a lamb producer, not a butcher. Our whole ethos was to produce a quality product for direct sale to our lovely customers and there was no point in abandoning it, simply to make more money.

So we pressed on, as ever trying to reduce our farming costs while finding other ways to add value to our retail sales.

The paperwork for the sheep was beginning to get quite time-consuming. With three to four hundred lambs a year going to the abattoir, plus around fifty older ewes for mutton, it was a complicated task to keep track of the records, both for the law and for our own management purposes. The new regulations required us to report movements of animals electronically to the delightfully acronymic ARAMS (Animal Reporting and Movement Service). This was to be done via an electronic link and it seemed a sensible time to revisit a computerised recording system.

This second attempt was, as it turned out, a step too far for our modest little operation.

I was wandering round the 'farm electronics' section of one of the local sheep shows, renewing my acquaintance with the software suppliers whom I knew well from my ADAS days. 'If you're farming now, you should be using one of these packages that you helped to develop,' grinned Matt, the manager of one of the longest established software houses. 'We'll do you a good deal for old times' sake!'

'Ah – but I've got the practical knowledge now,' I laughed, 'And I never worked on the sheep system but I do at least know what I'm looking for.'

'What do you need then? Individual records, health, drug use?'

'Yep, all of that, but the biggest problem all sheep farmers struggle with is lost tags. My old system couldn't link a new tag number to an old one and just treated a retagged sheep as a new animal.'

'Our new version does exactly what you want.' A warning message if ever there was one – as I'd been telling farmers for years. But I went along with it and bought a system complete with a very expensive new tag reader.

I never really got to grips with it properly. It was designed for big flocks where one action like a worming treatment was applied to many animals at once. In our case we wanted to do the reverse and record several things against a single animal, and that proved very cumbersome. It worked OK but with some weird logic that kept recording more animals going to the abattoir than I actually took. This took weeks to sort out and as the system automatically transferred this information to ARAMS, the mess it created took a while to unravel.

If you're interested, and I'm sure most of you aren't, the system had a 'weaning' function that separated the lambs from their mums in the database. And if the system wasn't told that the lambs had been weaned, then it decided, a bit like Bo Peep, that everywhere the ewe went, her lambs were sure to go. So moving a ewe to the abattoir at the end of her useful life without pressing the magic 'weaning' button meant the system thought that the lambs had gone too. Hence the chaos.

Our holy grail was to record lamb weights through the year and work out which ones grew fastest so we knew which ewes to keep and breed from. But we never managed it. Our weigh scale was an old manual one so we had to scan the ear tag and type in the weight of each lamb. It was just too time-consuming. Modern systems automatically scan tags and record weights and can calculate and display growth rates instantaneously. The really clever ones can even divert animals into different pens using automated 'shedding gates' and I have seen farmers set a weigh crate and shedder working and then leave the lambs to weigh and sort themselves! Sadly, we never achieved those levels of sophistication, but at least we tried.

Rather more successful was our second go at automated lamb feeding. Frances was not impressed. 'I sold all the lamb feeding equipment to stop you doing anything silly like rearing Tiddlers ever again. Why on earth do you want to do it a second time?'

'Well, you're the one who has to hold eight bottles in two hands when you're feeding the orphan lambs and then wash them all up afterwards. I figured we should try again – it'll save time and remember most of the problems we had last time were due to Orf. The key thing is that this feeder is different.' I sensed her scepticism.

'The old feeders we used for the tiddlers were just buckets with a heater in them so they stored the milk warm. I reckon that's why we got so many digestive upsets. This new feeder stores the milk cold and as the lambs suckle, the milk is drawn through a water bath and warmed up. I'm sure this will reduce the bugs in it and hopefully reduce the scouring (diarrhoea) problems that we had last time.'

'But it'll be endless cleaning again and I'll have to do it.'

'Ah but the students can do it now and it comes with a little pump that you (or rather they) can use to flush the pipes out with bleach. It's no worse than washing loads of bottles twice a day.'

She wasn't convinced but agreed we should have a chat with the supplier – who was a friend of ours. He persuaded us to try one of his prototype models – and it was an immediate success. The lambs grew ferociously fast with virtually no digestive problems. It was easy to clean and the students took it on with enthusiasm. We even got featured in the *Farmers' Weekly* as an early adopter of the new system and got a dinner for two out of it from the supplier as a thank you!

On reflection I think it worked better not just because the milk was stored cold but also because it didn't warm it excessively. I'm pretty sure that more digestive problems in lambs are caused by milk being too warm than too cold. Ironically, the only downside was that it was too easy. When struggling with a difficult ewe and a lamb reluctant to suckle late at night, it was always very tempting to just give up and put the lamb in the orphans pen!

ON THE RETAIL SIDE, we were still on the lookout for new ways to grow our income. We looked at box schemes as these seemed to be popular with other suppliers but I couldn't see how someone would be willing to take a box sufficiently regularly to make it work financially. It wasn't really that different to our Half Lamb boxes except that people would have to commit to a more regular purchase. I couldn't see how we could emulate the veg box schemes where customers seemed willing to accept whatever was in the box week after week. We tried out the idea on a few regulars but they all said they'd rather specify what they wanted or buy direct from the market stall on the day. I suspect we'd have been more successful if we'd had a better on-line presence, but I'm sure people would still want to specify what went into their box.

We also looked at a couple of on-line ordering schemes to try to expand our sales.

Typically these systems, involve a third party who provides a web site on which suppliers can list their produce. Customers place their orders online and the website provider amalgamates the orders and passes them to the supplier to fulfil. And unfortunately, that's where it usually goes wrong as it's the fulfilment that usually fails.

One scheme we looked at tried to organise hubs where produce could be left for customers to collect; another even tried using delivery vans for their shop equipment servicing business to deliver to local collection points. Sadly, none of these systems worked for us and all failed eventually. So apart from a few online sales from our own web site, fulfilled by expensive courier delivery, we pretty much focussed on face-to-face sales

OF COURSE INCOME didn't have to be generated directly from selling our produce. We had skills and facilities that we could exploit to earn money. Perhaps the most bizarre was the idea of team-building experience days for corporate customers. We organised a whole day including moving sheep into a pen, sorting and marking them, moving them using a Land Rover and trailer (driven by my son) and crutching and dosing them. They also had a simple butchery task and the day ended with them cooking their prepared lamb on our griddle for their tea. It was a huge success and very lucrative but it was a one-off for Ross's company. We did consider developing it but decided it would be too disruptive to our farming routine and take too much time and cash to promote it to other customers.

More realistic was the idea of B&B stays where the guests could help with the farm chores and learn about farming and food production. This worked up to a point: most people wanted to stay on a farm but weren't that interested in the 'experience' part of the offer. We kept the B&B going as it was a useful addition to our income, although bookings almost inevitably coincided with busy market weekends. This meant that we ended up paying people to come in and serve breakfast for us, which rather defeated the object.

We did, however, have one couple take up the farming experience – a rather strange American pair who called us from Heathrow airport saying they'd like to drop by for the night. Heathrow is about two and a half hours away! They could see us on the map and wanted to know how far away we were from Hereford – which they pronounced 'Here Ford'. They turned up and wanted a tour of the farm, which I duly provided, before they disappeared to find an evening meal. I left the following morning before they emerged, but when I got back Frances handed me a present they'd left me – a booklet about caring for our 'flock'. They were evangelical Christians on their way home from a missionary posting somewhere in Africa.

After these rather underwhelming attempts to increase our income, we decided to look a little closer to home. I had skills

*'I could charge a fee for the demo and then sell
the butchered meat at the end.'*

that I'd developed that were useful, most notably in the butchery. I was no master butcher but I could cut up lambs very
effectively and had developed some imaginative cuts that made
the most of our produce. So we decided to capitalise on my
butchery skills.

I was already doing evening talks about the farm, mostly
to Women's Institutes and other rural groups and it was relatively simple to add a Butchery Demo as a second option. This
worked very well as I could charge a fee for the demo and then
sell the butchered meat at the end of the evening. It was also
great fun and I was often surprised at how little these country
folk knew about the food they were eating. We had one difficult meeting where two ladies walked out as they were vegetarians. The WI president summed up everyone else's feelings
with the acerbic remark, 'One wonders why they came?'

Our other idea was to offer butchery training days where
one or two people could come to the farm to learn how to cut
up their own lamb. This was a very popular offer and a very

good earner. I could charge a course fee per head for the day and each participant would buy a lamb to cut up. Our only cost apart from my time was lunch for the participants which was unsurprisingly, a lamb dish.

ALL THESE INITIATIVES were helpful but we were still a long way from our aim of a business that we could step back from. We weren't even in a position to employ casual help apart from the few hours that Barry did in the butchery. What had been fun in the early days was becoming tiresome and wearying. With just the two of us, the retailing routine in particular was becoming relentless. We were experiencing a little existential angst.

The wind was howling in the Yew tree outside our bedroom window. I rolled over to peer at the clock – it was four a.m. Still an hour and a half before getting-up time but I was wide awake. I tried to blank my mind of the hundreds of things popping in, essential tasks I had to do in the next few days, but the harder I tried to forget, the more I remembered.

Orders for the weekend, students for lambing, ordering feed, how much lamb to keep over for last-minute Christmas customers – would the same woman call again on Boxing Day as family descended, apparently unannounced, demanding a leg of lamb to roast? Everything crowded in as fast as I tried to push it out again. Then, just as I was starting to doze, I heard a crash. The wind had dislodged one of the dustbin lids and sent it rolling across the yard. I traced its path in my head, waiting for it to fall on its side like a top. And it did, with a hollow metallic thud.

I put the radio on and listened to the World Service until the changeover to Radio 4 at 5.20. I then followed the familiar pattern of the shipping forecast, knowing that when it got to Ardnamurchan Point, it was time to get up.

It was a market day and our last before Christmas, so we

had lots extra to do. It was one of those 'double-headers', as we called them, where Frances went one way (to Hereford on this occasion) and I went another – to Ludlow today. Two vehicles to load, two lots of produce to pack, two lots of Christmas orders to send to the right place. I had at least had the sense to sort the meat for the two markets into separate trays last night, I thought, as I stumbled into our freezing bathroom. This was a recent innovation – the packing not the bathroom – that avoided the heated negotiation in the cold room over who should take what produce to where. And while it might sound incongruous, believe me, you can have a very heated conversation in a cold room at six o'clock in the morning!

We, or to be more accurate I, lived in fear of taking the wrong orders to the wrong market on days like this, but we never had. 'But there's always a first time!' I mumbled as I made my way downstairs into the warmer kitchen.

With the prospect of seven hours in the cold, we always tried to have a decent breakfast on market days, but I always used the cold as an excuse to try food from friends on the other stalls.

Getting stalls set up in Hereford was a herculean task which the stall-holders had agreed to do themselves, in return for a better deal on stall fees. This worked well in the summer but on cold, wet and windy mornings in the dark, it wasn't quite so much fun. Luckily I wasn't going to Hereford today (and Frances wasn't expected to do the heavy lifting) and Ludlow had permanent stalls that were pretty much wind-proof. They were, however, in the market square at the highest point in the town and surely one of the coldest places on the planet.

I got set up and parked the truck. The market was coming alive as it got light, everyone keen to make the most of the Christmas trade. I treated myself to a coffee from a little café just off the square and as I turned to make my first sale of the day, a gust of wind tipped the whole cup into the chiller. My lovely produce with its distinctive green labels all turned a dull grey colour and, with a muttered oath, I grabbed the sodden packs of chops and steaks and wiped them off as carefully as I could. I managed to soak up the coffee in the chiller and meanwhile my first customer stood patiently while this little pantomime played out, mercifully refraining from laughing. She took her orders and with a hint of a smile said, 'Good luck, I think you might need it today!'

The rest of the day was uneventful by comparison. The wind got under the stall roofs once or twice and tipped the rain water that collects on them down our customers' necks – not great for customer relations. We tried to remove the water by pushing up the sheets from underneath with a broom handle, but it was raining too hard by then as well as blowing a gale.

It was still raining several hours later as I backed the truck up to the packing room to unload. I'd had a good day and the disheartening job of relabelling the unsold meat for the freezer took less time than usual

It was dark by now and the lights from the house lit up the yard as I went in for a quick coffee and a warm-up. We often didn't see the farm in daylight on market days and I still had a couple of hours of farm work to do before I could get warm properly. But at least we had a cosy evening by the fire to look forward to.

Our neighbour Bert worked many more hours than us before he retired, rising before five a.m. every day, not just on market days. He worked alone, in all weathers, and he never seemed to dress differently except in the height of summer when he might shed his tattered old jacket for a few hours. As I stepped outside again, I noticed a few snowflakes swirling in the yard lights. I pulled up my coat collar and trudged up the lane towards the barn where the sheep were waiting for their tea.

The wind was still whistling through the hedges and the

rain had given way to large flakes of snow. 'That's all we need,' I muttered as I wondered for the umpteenth time why we did this. Unlike Bert, we didn't really need to. It was a choice we'd made, a dream, a goal or whatever you'd call it, and in the summer it was rewarding and great fun. But in these conditions, it was just damned hard work.

Through the gloom and the swirling snow, a Land Rover appeared with Bert at the wheel, off to check his tenant's sheep. He stopped. 'You OK Andrew?' he laughed. 'It's a bit fresh, isn't it, but this'll all be gone by mornin' – it'll turn back to rain.'

I marvelled at the resilience of chaps like Bert. They'd spent their lives in these conditions, they accepted their lot and just got on with it. I really had very little to worry about: a cosy home, a wife who would have something ready to eat when I got in, and a future that didn't depend on me farming sheep. Bert had none of these, but he still kept smiling.

I was beginning to realise that we needed to get past this tiresome, relentless stage if we were to succeed. There wasn't much we could do about the farming; we couldn't afford to rent more land or buy more sheep and we'd taken out as much cost as we could, so we were left with only one option. We needed a step change in our income, something really big that would make a difference.

7. Moving Up a Gear

A GLOOMY WINTER'S AFTERNOON in Hereford High Town wasn't the most likely venue for a business strategy breakthrough. But by then, our customers had all gone home so we'd plenty of time to chat.

Chris pulled up his collar against the bitter cold. 'I'll be glad when I've had enough of this,' he mumbled through his scarf.

'At least you've sold most of your stuff,' I grumbled. 'I'd have thought this weather would have encouraged folk to think about roast lamb dinners. But they clearly have other ideas!'

'This is bacon-and-sausage-sarnie weather – who wants the faff of roasting a joint of lamb on a Thursday? Your 'mid-week roasting joints' are just a gimmick – it's time you started selling proper food!'

'Any fool can sell sausages,' I laughed, 'they're an easy sell, an impulse buy. I'm going for the more discerning customer.'

'So discerning that they're sitting at home in the warm by the fire,' he smiled, 'the ones not daft enough to come out on a miserable, February afternoon like this?'

He was right, of course. Behind the good-natured banter was a shrewd business brain.

'Now we're doing these markets weekly, what are you going to do with all the left-over meat? If you don't do something quickly, you're going to get buried under unsold lamb.'

'Freeze it, like we always do, and sell it direct at the door,' I ventured rather lamely.

'And when your freezers are full?'

'You have a point,' I conceded. 'So what do you do? And don't tell me you sell out every week because I know you don't!'

'Carcase balancing is a big deal in both our games,' he said, warming to his theme, 'and now you've moved away from selling whole and half lambs, you keep getting left with the less popular cuts. And then there are days like today when you're taking half the stuff home again. We've found that processing the meat into sausages gives us options – it's more flexible.'

'We couldn't make lamb sausages though, could we? How would that work?'

'I doubt it would,' he replied. 'We can get sixty kilos of sausage meat from one animal – what would you get from a lamb?'

'Fifteen kilos if we were lucky.'

'So you'd have to charge a fortune for them and you'd be back with the same problem of a high-priced product and a difficult sell.'

Chris was in his stride now. 'Have you thought about burgers?' he said. 'You could freeze them and sell them either as a product to take away or maybe even cook them and sell them as 'food to go'.'

'We could mince unsold meat after each market and make

them to freeze and we could probably thaw out frozen meat and make burgers to cook, as long as we didn't re-freeze them?'

'I'd think about it carefully,' cautioned Chris. 'It'd be a lot of work and you'd need a proper mincer and burger former. That cheap thing you use to make a few packs of mince each week will give up very quickly. But I reckon it's worth a look.'

'It would certainly get us into the food vendor market,' I said, mostly to myself. 'We could get into the festivals, selling food and make some real money!'

'Go and do the sums,' laughed Chris, 'but processing is where the money is.'

The light from the shop windows shone on the wet flags; High Town was deserted.

'It's time to go.' I was cold now as well. 'That's enough business planning for one day.'

'My consulting invoice will be in the post,' smiled Chris, 'and I'm even colder now having sorted your business for you.'

I walked to the truck only half-listening to Chris's chatter. I was deep in thought. As so often happens, a chance conversation with a friend and fellow trader had just possibly provided us with the next crucial step in our retailing career.

NEITHER FRANCES NOR I particularly liked burgers. They were reputedly made with cheap off-cuts and contained all sorts of nondescript nasties and lots of fat, or worse. But some trader friends of ours were selling a beef burger made from beef from their own farm under the slogan 'rehabilitate the burger', which basically meant making them properly, with good ingredients, and focusing on quality. We thought we could do the same with lamb but we had no idea where to start.

'I know how to make burgers,' enthused our friend Hazel. 'It's easy – just mince the lamb, add some herbs and bind it with egg.'

Other friends suggested different flavourings as they did for their children and one even came up with a plan to serve them with salad and mayonnaise in a ciabatta roll.

But we'd been here before: enthusiastic and well-meaning friends giving us advice. We needed commercial burger production know-how. We knew we couldn't use eggs – our Environmental Health Officer would have a fit. And using herbs and other fresh ingredients would be great but how would we keep the product consistent? Eventually we found a supplier of burger mix on the internet and ordered a trial pack. It contained rice flour and flavourings and there was a recipe with it. We then went out and bought a tiny plastic burger press, like our local friends used, and set about our first venture into meat processing.

We had a lamb shoulder left over one evening after a butchery session and I decided to bone and mince it for a trial burger run. Our little mincer coughed and struggled and we had to keep removing the cutting discs and cleaning them but eventually after what seemed like for ever, we had a pile of mince ready to mix. We found a bowl and measured the burger mix carefully in an old butter carton.

The mince and the burger mix was then re-minced as the recipe instructed, which proved very difficult and time-consuming as the mix refused to move freely through the mincer. We settled on six-ounce burgers (150 grams) for no other reason than it seemed like a decent size, and carefully formed them with our little press. The final step in this rather chaotic and ad hoc process was to freeze a dozen burgers on a tray (as Chris did with his sausages) so they didn't stick together.

The next day, we took them to the market in Leominster and offered them as free samples for some of our regulars to take away and try, with the proviso that they call us and report back.

And they did: within days we had huge enthusiasm and demands for more. And so started the 'next big thing' in Whyle House Lamb's journey.

As Chris predicted, our domestic mincer burned out within a few weeks and the plastic press was too small to handle the volume we needed, so we decided to go to a trade food show in Birmingham to find something more appropriate.

The food world is very different from the farming world and we wandered round that show like a couple of bumpkins, way out of our depth. Someone tried to sell us a burger maker for a reduced price of £12,000, and everywhere we looked it was about scale and speed rather than quality and provenance. We did however find the supplier of the sample pack of burger mix who proved very helpful and gave us lots of advice on mincing, mixing and forming burgers on the scale that we were aiming for. We also bought a new burger press, a proper one that formed beautiful burgers time after time, albeit one at a time. We had discussed automated burger-makers with a catering friend who told us to steer clear: 'They cost a fortune' – which we already knew – 'and the meat needs to be like slurry for them to work properly' – which we didn't – was his opinion. He underlined this advice by stating adamantly that he only bought burgers from butchers who pressed by hand!

We also bought a second-hand professional mincer to be delivered the following week and lots of trays and a huge mixing bowl to handle bigger batches.

On our way home, we realised that we were now committed; we had the kit and we'd spent the cash – we had at least to give it a go.

Our new mincer worked a treat. It chomped its way effortlessly through all we fed into it and we found we needed to mix up the cuts to keep a reasonable fat content in the mince. Lamb breast and shoulders were quite fatty but loins and legs were too lean on their own. We had a friend who wanted spe-

184

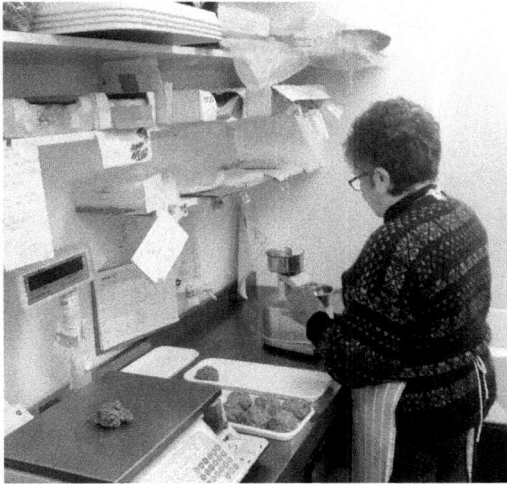

'Sunday very quickly became 'Burgering day'.'

cial low-fat burgers made with loin and leg and we thought this would be a great extra line. But they were too slippery and wouldn't form properly – another great idea destined not to fly.

A butcher friend of ours wasn't impressed. 'What on earth are you doing putting legs and loins into burgers?' he demanded. 'You should be getting rid of your cheap cuts in these.' Which of course is exactly why burgers have such a bad name. I once worked out that a prime leg of lamb worth £25 at the time was worth £29 as burgers, so the calculation was always in the right direction, even for the expensive cuts.

Our burger-making enterprise grew very quickly, using unsold meat from the markets to make burgers that we could freeze and sell. Sunday very quickly became 'Burgering day' so we could clear the cold room ready for the new week.

We could sell packs of burgers on the market stalls very easily and we quickly realised that offering tasters brought people to the stall who would then buy burgers and, usually, other cuts as well. This involved yet another 'bag' to take to markets with a cooker, pan and oil, but it was worth it and occasionally we'd have a few bread rolls and sell some as 'food to go'.

A lesson we learned very early was that retailing is largely about pack price. Our initial offering was four 150-gram burgers with a pack price of £6.75, which was too much. We changed to four 100-gram burgers with a pack price of £4.50 and they flew off the stall. The same price per kilo but a pack price of under £5 was apparently far cheaper!

AS A ROUGH RULE of thumb, a lamb sold as joints and chops on a market stall just about doubled its live value; sold as burgers in packs to take away trebled it, and sold as a burger in a bun, more than quadrupled it – in other words an £80 lamb would be worth getting on for £350 sold as burgers at a festival. So at this point we decided to take another look at festival trading.

'I've filled in the forms and offered a decent price for Ludlow Festival, but I've heard nothing,' I said to the ever-helpful Chris. 'I'm really not sure this festival thing is for us – it's too stressful!'

'Just give them time,' he said calmly. 'You're a new trader, they'll want to suss you out, but you'll get in. And in any case, with the markup you were talking about, it's a no-brainer.'

Ludlow festival in those days was organised by our friend Beth. A mixture of ruthless business acumen and complete lunacy that endeared her to traders and customers alike. I wasn't entirely sure why I should plead with an organiser to be allowed to pay them to trade, but that's the way it was (and still is) in the festival world. I knew they had to balance getting a good range of stalls to attract the punters with being fair to their existing traders, but it required a degree of persistence to get accepted. This was our third year of trying at Ludlow and I'd been careful to get to know Beth and assure her that we were

'I even called her our 'Lamb Burger Ambassador'.'

doing something special and new. For later festivals, I also had another ploy up my sleeve, which was to ply her daughter, Jess, with free lamb burgers at every festival. I even called her our 'Lamb Burger Ambassador'. I'd re-written the proposal with new menus and prices and lots of 'local' references to make it more appealing. But there was only so much creativity I could come up with – burgers were burgers, after all.

Eventually we heard that we'd been successful. 'Blimey! We've got a bill for £700,' Frances called from the office one morning. 'It's only May, Ludlow's not for another four months!' It was a shock to our cashflow and a big risk as we wouldn't know until the day before if we'd got a good pitch or were to be stuck behind a wall somewhere out of sight. This was why I'd been so wary of festivals.

I decided to ask others for some advice and was warned that selling food-to-go at festivals required a different mindset to the more sedate farmers' markets: 'Your aim is to feed as many people as you can as fast as you can and you need good equipment, trained and motivated staff, and extremely good organisation.' So we bought a proper electric catering griddle and persuaded family to come and help us – the organisation part would have to be learned on the job.

As the festival approached, I realised I had no idea how much produce we'd need to take. It was held over three days but the Friday was a 'trade day' so that'd be quiet and Frances could do that on her own while I prepared for the weekend. Eventually I worked out (or more truthfully, guessed) that we'd need fifteen lambs that week, bearing in mind we usually killed two or three at a time. This required lots of weighing and sorting to find enough animals two weeks ahead so they could hang in the cold room for the required time. And it occurred to me during one of several sleepless nights that the cold room might not be big enough...

Ten days before the festival, I collected the lambs from the abattoir – and we couldn't get them all into the car. The slaughtermen thought this was a huge joke and fetched a large plastic bag in which to envelop the fifteenth lamb so they could sit it on the front seat with a seat belt around it. I got some strange looks from fellow drivers when I pulled up at traffic lights on the way home.

The week before was frantic, cutting up meat, trying to guess which cuts would sell and of course, making our burgers. As well as making them with fresh meat, we decided to try to reduce our stock of frozen meat by thawing some out to make what we called 'Festival Burgers'. These were special ones, used only for events where they were cooked under our supervision and sold as food to go. We couldn't risk selling them as an uncooked product as we couldn't be sure that folk wouldn't refreeze them. Environmental Health rules are very strict on this issue as refreezing meat can multiply the bacterial contamination exponentially. So we only made as many as we knew we could sell; those left over had to be recycled through the dog.

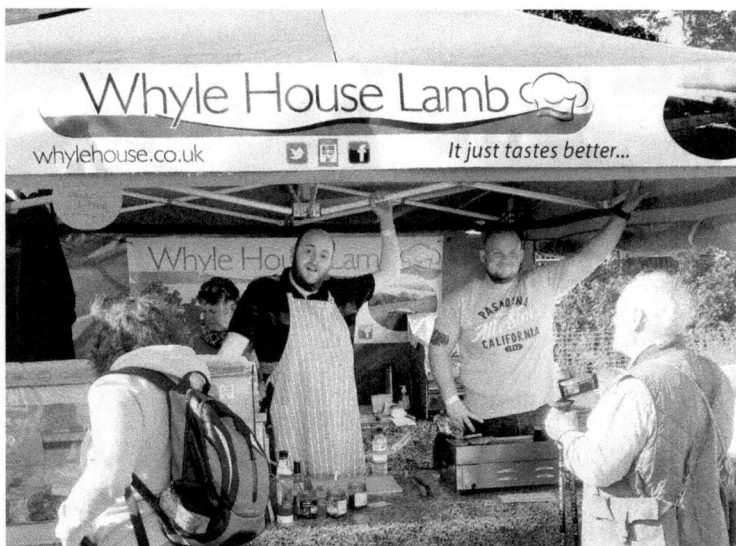

'We had a good pitch!'

As we were still finding our way, on this occasion we decided to take just a few trays of Festival Burgers and lots of frozen ones which we could sell or cook. We bought more bread rolls than we thought possible and realised, too late, that they wouldn't all fit on the back seat of the car.

On the Thursday afternoon, we made our way timidly to Ludlow Castle to set up. Everyone knew everyone else and we felt very much the outsiders. We drove carefully through the Castle Gates, which the medieval castle designers – rather shortsightedly, in our view – had made far too narrow, and got directed by a very helpful steward to our pitch. The instructions we'd been sent insisted that we gave the steward a twenty-pound note, which we would forfeit if we were on site for more than an hour, but he waived it away with a smile. 'Nobody does that now.'

We had a good pitch! Right by the main display area and where folk tended to hang out later in the day and picnic with their friends. We got the stall set up, unloaded our gear and hurried back to the butchery to continue our prep. We decided to focus on produce and burgers to sell for the Friday when Frances would be on her own. We could go for the food-to-go trade over the weekend when we had more help.

We worked late into the night getting ready and crawled out of bed at six am the next morning to load up and make our way back to Ludlow. I felt a bit guilty about leaving her on her own but she assured me she'd be OK and in any case, with a mountain of work still to do, I needed to get home. I was part way through a batch of burger pressing when I got a call; she'd sold out already and it was only three o'clock.

'There's no point in bringing extra today or we'll not have enough for the weekend,' I said. 'Get out as soon as they'll let you and we'll crack on with the weekend stuff.'

With dire warnings ringing in our ears about the queues getting through the Castle Gates on Saturday mornings, we arrived at seven a.m. to a deserted site. 'Blimey, couldn't you sleep!' laughed one of the stewards. 'Get in and out as soon as you can though.'

I did as I was told, parked the truck and made the long walk back up the hill from the traders' car park; a walk I was destined to do many times in the coming years.

The castle grounds were busy when I got back, cars being unloaded, trucks backing through the huge marquee doors and people scurrying everywhere. Most were in a good mood because a) it wasn't raining and b) they'd had a good day on Friday. We'd heard tales of wet festivals when all the produce had to be unloaded outside on the square and moved into the castle by hand trolley – an experience we were only to have once in all our festival years.

We were also warned of problems with the electricity supply, which happened at most festivals in most years. I decided to make friends with the electricians, and bribe them with the promise of free lamb burgers later in the day. This proved to be a very good move and Dave and Fred did indeed become friends whom we saw at most festivals in the West Midlands. They always sorted problems with a smile and even, on one occasion, fixed a meat chiller on site for us.

As the ten a.m. opening time approached, I turned on the griddle and put a couple of burgers on for tasters. Son Ross and his friend Lewis arrived to help. I tried to explain to them how it was all going to work. They just smiled. I could see them thinking, 'Silly old fool, it's only a few burgers for goodness sake.' Or words to that effect.

At ten sharp, the first punters arrived; we were offering tasters on cocktail sticks to encourage them to come back later for lunch. 'Can't I have one now?' asked the second one in the queue. I gave him my other taster burger and hurriedly put on more to cook. 'Looks like I'd better cut some rolls,' said Ross, taking charge. 'Get Lewis cooking and you take the money and sell the meat.'

We settled into a routine and served burgers to a steady stream of people all morning. As lunch time approached, our queue grew longer and we increased the pace. Rolls were cut, placed in a napkin and handed to Lewis the right way round so he could put a burger straight in and pass to the customer. Ross and I were both taking orders but I was taking the money so I didn't touch food to keep the health inspectors happy. We had a big container of hand sanitiser on the stall – as ever, trying to be the good guys – but it looked a bit like a ketchup dispenser. After the second customer sprayed his burger with sanitising foam, we ditched it – people could take responsibility for their own hygiene.

We stepped up the pace again – frenzied now – and kept up a constant banter with those in the queue, trying to keep them there until we could serve them. We didn't lose many.

As the lunch trade slowed, we realised two things. We had no change left, and we were going to run out of stock. An anxious call to Frances got a promise to bring more burgers as soon as she could.

Daughter Hannah arrived at three p.m. with more stock and we carried on serving at a manic pace with a twenty-deep queue, at least half of whom were coming back for more. Many had had plenty to drink and wanted to tell us, and everyone else, how good they thought our burgers were. Ross gently moved them on, keeping up a constant banter while Lewis and I continued serving. By six o'clock, the gates were closed and most had left, apart from the over-excited ones who kept coming back for more until we'd none left. We finally got packed up and, as Ross and Lewis left for home, I stumbled wearily to fetch the truck to take home all the empty boxes, ready for the next day.

At home, Hannah was busy pressing more burgers and Frances had gone off on a bread roll hunt. We'd used all our supplies for the whole weekend and we still had a day to go. I decided to butcher some of the next week's lambs and make more fresh burgers – which kept me out in the butchery until after midnight.

The next day, we did it all again. Except there was no Ross and Lewis and no Hannah back at base. It was just Frances and me. If anything, it was busier than the previous day and we struggled to keep up. Frances refused to accept that rolls had to be handed to me the right way round to put the burger in, but she was better at handling change than me so we didn't run out so quickly. I made a mental note to plan change based on prices – burgers at £4.80 mean lots of 20ps are needed – and lots of fivers.

By four o'clock we'd sold out and had to turn disappointed customers away. They were very good about it but it was so frustrating.

That evening, after a long and weary clear-up, we laid the money out on the kitchen table to cash up. I'd never seen so much cash in one place, ever. We counted it several times and I just sat and stared at it. Just about ten percent of our total annual turnover, sitting on that table, earned in just three days.

We did many festivals over the years but this was our baptism of fire; we learned to plan ahead, to be organised and to have plenty of help. We always had plenty of burgers as we could take frozen ones as a reserve, but our insistence that we always used fresh bread rolls meant that we usually ran out of those first. Initially, we took burgers as an adjunct to our meat sales but very quickly realised that, important though the meat was, it was the burgers that made the real money.

I occasionally did a festival butchery demo in later years and this proved a major logistical challenge – getting a lamb carcase and my kit onto site and then onto the demo stage at the right time. But I always returned to the stall to find my audience from the demonstration in the queue to buy our meat.

WE HAD STUMBLED ON a way to make good money from our lamb. At last we had a reliable source of large chunks of income that would make a sizable dent in the overdraft several times a year.

We quickly found however that we couldn't cook fast enough at the big festivals and decided we needed a bigger griddle. Friends Chris and Guy had a proper gas one which was very efficient and very cheap to run. But it was expensive to buy and difficult to transport. Ever resourceful, some would say cheapskate, I decided to modify our gas BBQ to work on a table top so we could transport it easily and augment our cooking capability.

I got it all tested and approved and it made its debut at a Ludlow event the following year. All was going well, I was cooking on it while Frances and her helper that day were using the electric griddle and we were serving a huge queue very quickly. It was also quite a spectacle – with lots of sizzling and BBQ smoke and smell. Then suddenly, there was a roar and flames shot upwards. The thing had caught fire: the volumes of cooking meant that the fat disposal mechanism of a domestic BBQ just couldn't cope – and we had flames and billowing black smoke for several minutes. Luckily, I'd heeded the rules that insisted that gas cooking should only take place outside the covered area of the gazebo! I managed to damp down the flames and remove the gas bottles to a safe distance - and received a cheer and applause from the queue who weren't the least bit concerned. I'm not sure the organisers would have been so accommodating but thankfully they didn't see it. At this point, we decided to buy a second electric griddle.

We were very appreciative of all the family help at the festivals. It saved us a lot of money and I like to think they enjoyed themselves but they weren't above being critical of our efforts. Or at least Ross wasn't.

We were at the Ludlow Spring Festival – like the autumn one but with vintage cars and beer. Ross had been telling me for years that we were missing a trick, selling just burgers. Like most youngsters he had little regard for his father's judgement and I'd ignored him, but the unexpected fine weather had brought customers out in droves. Our burger stocks were going down alarmingly quickly and we had no backup at home to bring extra supplies, so I finally agreed to cut up some of the lamb steak to cook on the griddle and sell.

Despite the very rough butchery, with a knife borrowed from the ever-supportive Chris, these were an instant success. We continued all afternoon, augmenting our burger sales with Lamb Steak Rolls made by cutting up our beautifully butchered and packed meat from the chiller. By seven p.m., when the

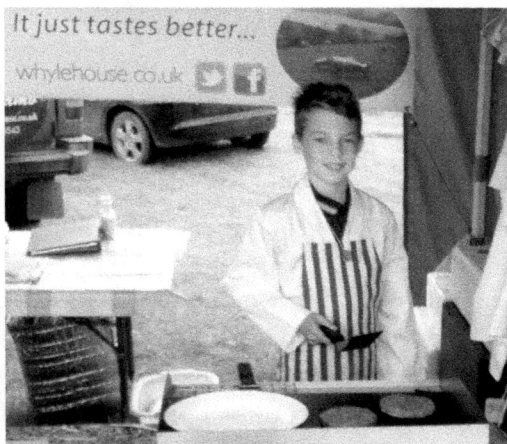

'Ollie piped up, 'Give me a plate Grandad,
I'll take some tasters round the site.'

organisers asked us if we'd stay for a while as the fine weather meant the site was still full, we had nothing left to sell. All the joints, chops and steaks had been cut up and cooked and every burger had gone.

That was the only time we walked off site with no left-over product at all – nothing – and although another night in the butchery beckoned, it was a great feeling. We'd also discovered another profitable product line – almost by accident – although Ross would of course claim otherwise.

MY FAVOURITE FESTIVAL was Shobdon, the first in the annual calendar, and it was always great to see our friends after the winter. On this occasion we had a great pitch and the organisers – Matt Teale, the newscaster of Midlands TV fame, and his affable father, Terry – were running their usual relaxed but very tight ship. I had grandson Ollie with me, as he was eight by then and liked to help at the smaller events. We'd even got him a small white coat so he looked the part. Trade was a little slow that morning, it'd been raining and people were delaying their arrival, when Ollie piped up, 'Give me a plate Grandad, I'll take some tasters round the site.'

And with that he picked up a plate, removed two burgers from the griddle and cut them into pieces, skewering each with a cocktail stick. Before I could say 'Be careful!' he was off. I watched him for a while as he marched up to groups of people, offering his samples. They all smiled as he proudly presented his plate and then I lost sight of him as the site filled up. He returned later for more supplies and disappeared again. After a while I realised I'd got a queue building. 'Ah that's better,' I smiled, 'trade's picking up, I think people are coming in now.'

'No, we've been here a while,' said the queue leader. 'Ollie's told us to come and buy his grandad's burgers before you sell out!' At that point, Ollie returned with his empty plate and the customer gave him a tip, which he put in the cash box.

'Don't you do that,' he laughed. Ollie looked startled. 'That's for you! You're a grand little salesman and your grandad's got plenty!'

As I implied earlier, there is something of an art to getting accepted to trade at festivals. You can't just turn up, and even well-established traders have to wait their turn. We spent many years trying to get into the Ludlow Medieval Fayre and finally succeeded with our new product line of 'Medieval Mutton' burgers.

Following our usual theme, this had happened by accident. One evening I had finished preparing mutton for a market the next day and had a little left over. Frances had been making lamb burgers and had all the equipment in use so we made a few mutton burgers – with the same recipe as the lamb but using minced mutton instead. We tried them on the market the next day and sold out within half an hour.

The Medieval Fayre was run by the mother and daughter team of Prue and Abbie – lovely but quite fearsome ladies who

'We arrived on site with warm clothes under our costumes, looking a bit like Teletubby knights.'

ruled with a rod of iron and insisted that everyone dress in costume. Dressing up was a pain, but it did add to the atmosphere, and when the weather was good, we did very well there. We invariably needed help at this one, usually Frances's daughter-in-law, Celine, who always did a great job and was very popular with the customers. She detested dressing up even more than we did, but she had a competitive streak that meant she and Ross were always arguing good naturedly over who could take the most money.

On this particular occasion, we arrived on site with warm clothes under our costumes, looking a bit like Teletubby knights. We'd already been on site twice: once to set up the stall, and again to move it as it was 'too far forward'. With anyone else, I'd have probably argued, but I knew it was pointless with Prue and Abbie.

Celine joined us and reluctantly donned her costume. This was going to be a good one. With lots of entertainment and continuous shows on the stage opposite, we were at the focal point of the event, and armies of people would be wanting food. The shows repeated every hour or so and we knew that, as usual, we'd have all the routines word-perfect by the end of the day. The lunch-time surge came and went and we marvelled that even though we told people – every time – to come early or late to avoid the rush, they never did.

The younger grandchildren arrived in the afternoon and dashed off to spend their money. Occasionally they came back to try to help, but it was still frantically busy and we didn't have time to supervise them. As dusk fell, the crowds grew, waiting for the pyrotechnics show from within the castle walls. Apparently it was a truly spectacular finale to the first day but we never saw it as we were always so busy.

This day of the Fayre was always a challenge as we traded into the evening. Getting enough stock on site and persuading our unpaid helpers to stay was difficult. But on this occasion, we were busy right up to the close at nine p.m. and although it was a long and tiring day, we achieved our highest-ever daily take. That's a record that Ross still disputes, arguing that trading into the evening gave Celine an unfair advantage. Kids!

Getting off site in the dark was even more difficult than for the summer festivals, but we were wise now. We'd worked out that we could remove unsold produce on a hand trolley to the car park and come back the next morning to clear up. So while our friends were still queuing to get their vehicles through the castle gates, we were at home, with a glass of wine, in front of the fire!

Like Prue and Abbie, most festival organisers ruled with absolute power and we, as mere traders, did as we were told. But it wasn't always quite so one-sided.

We were at Cosford Festival this time and we were all looking glumly at each other. It'd been a disappointing day at what had always been a really excellent show for everyone.

The event was in its third year, organised by the indefatigable Abbie (another Abbie) but in an attempt to diversify the food offering, they'd invited too many traders selling 'food to go'. So none of us had made money that day. This had happened some time previously at another festival and we'd given firm feedback that we wouldn't attend if they did it again. It worked there so we tried again here and sure enough, we arrived the next year to a diverse but smaller food offering where we could all sell enough to make money. Just occasionally traders could influence things, especially once they were established and popular with the customers.

The only downside to festivals was that they were much less sedate than your average Farmers' Market. The big ones could get quite boisterous, especially when there was alcohol around, and although the bulk of the customers were still lovely, there were those who were just determined to find fault. Determined to be offended.

'I was told I'd get a discount but I didn't, the burger was smaller than the others and the whole thing was a miserable experience' – a Facebook post this time.

We were at Ludlow again, frantically busy, but we'd learned that our burgers were attractive to those who were gluten-intolerant, so we always put 'gluten free' on one of the advertising boards on the stall. The burgers themselves were gluten free but we couldn't offer gluten-free rolls as we couldn't manage the cross-contamination risk on a busy stall. So we offered a burger on its own on a disposable dish with a disposable wooden fork.

Most customers were delighted with this option, but one clearly wasn't. She showed up at our busiest time and asked me lots of detailed questions about our recipe. The serious coeliacs never do that; they just quietly check the ingredients list and make a decision. After a lengthy debate, I moved on to serve a waiting queue and she turned her attention to Celine who served her. And, apparently happy, she walked away. A couple of hours later, a friend passing by the stall suggested I look at the business's Facebook page and there she was, in full tirade. The customer is always right, of course, so I offered to refund her money if she was still on site or by Bacs if she sent me her details. Unsurprisingly, we heard no more from her – some people just love to be seen to be offended.

The following year we had a much more serious attack, this time from militant vegans. I was preparing new season mutton and we decided to make a short video to promote it. It was well-received by friends and customers, so we made a series of them, which the vegans got to see.

'I won't block you,' I said naively. 'Let's have a discussion and see if we can better understand each other.'

I brought the world down on my head with vegans from all over the world screaming their invective at me.

'You're cutting up baby body parts; you're the son of Satan,' were some of the more printable tirades.

I tried to explain about animal welfare standards and abattoir regulations, but it was hopeless. I was quite shaken by how personally abusive it became and it took me a couple of weeks to damp it all down again – a lesson learned the hard way.

FESTIVALS WERE HARD work and took a lot of organising but the real challenge was fitting the demands of the farming around a retail operation that was becoming more and more time-consuming. We tried hard not to cut corners and for the most part succeeded. But one incident highlighted the tension between the two sides of the business.

I was rushing around preparing for a late summer event and had managed to squeeze an hour or so to check the sheep before it got dark. I drove onto our rented grass at Leysters, just as dusk was falling, hoping I could have a quick run round some fattening lambs and then get back to the butchery. But of course it wasn't that simple. These youngsters were our 'medium' lambs, not the real high flyers but those that would be ready through the autumn and early winter. They'd done well on some decent grass which made it doubly disappointing to see them starting to look unwell. They'd been fine a couple of days earlier, but they were now looking a bit lacklustre, and there were a few tell-tale mucky bottoms. It looked like the start of a worm infection but I didn't have the time to take faecal samples, let alone get them in and treat them. I reasoned that they'd only just started to look off-colour so they could wait until the following week.

I got back to them after the weekend and they'd lost condition alarmingly. I don't think I'd ever seen a group of sheep go downhill quite so quickly. Most had mucky backsides and they were clearly quite unwell. I got a few into the shed to look at and decided to bring them all home for treatment and some extra feed to get them back on track.

Harriet the vet was not impressed. 'They've really gone backwards,' she said. 'That's going to cost you, not just in treatment but in the feed you'll need to put them right.'

They recovered with no permanent damage done but it was a salutary lesson. We had finally found a way to secure the business financially but we needed to keep control of the farming at the same time.

8. WE'VE MADE IT!

PERHAPS THE MOST REWARDING show we did over the years was the biennial NSA Sheep Event at Malvern in Worcestershire. Organised by the National Sheep Association (NSA), it's a huge showcase for the industry where all the great and the good gather to talk and view all things sheep. We had attended for several years as punters and I even spoke at one of their seminars one year, but we'd never traded there.

Catering at large shows is not for the faint-hearted. Once you're selling food, you're in competition with the big catering companies who charge exorbitant prices for what I can only describe as poor quality food, generally served by underpaid, junior staff. As a result, they can bid very high pitch prices way out of reach of small businesses like ours. In my experience, they are also ruthless, competitive people who can be difficult to deal with. For this reason, we tended to stick with proper food festivals where the ethos was more about local, quality food than simply feeding masses of people. But the sheep event was the exception.

We attended as a 'local NSA member' doing our bit for local food and demonstrating a potential diversification. We also did the evening BBQ for the NSA staff and provided a plate of cooked lamb for the 'Love Lamb' promotion at the show. This promotion gave us a wonderful photo opportunity to have our lamb on display with two of the NSA Young Ambassadors, Fred Love and Hannah Jackson who is now known to many as 'The Red Shepherdess'. The catering companies weren't happy but 'allowed' us on site on the basis that we were so tiny we wouldn't damage their sales.

On our first day we were so successful that we ran out of rolls by lunch time. A kind soul on the next stall went into Malvern and ransacked all the bakers he could find to replenish our supplies. While he was away people kept arriving at the stall with an empty roll asking for 'just a burger.' After the fourth one, my curiosity got the better of me and I asked them

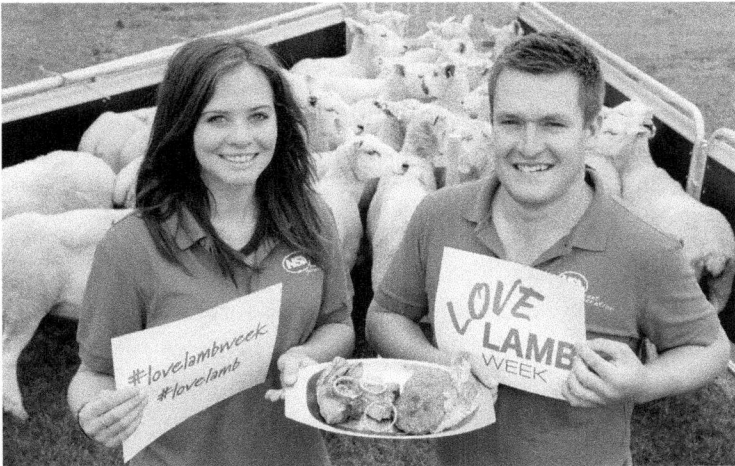

'Our lamb was on display with two of the NSA Young Ambassadors, Hannah Jackson and Fred Love'

where they were getting them. It turned out they were buying cheap burgers from the caterers, throwing away the meat patty and coming to us for a proper one...

An extraordinary event where we made a lot of money but the real reward was that we were selling to our friends and colleagues – our fellow farmers – who were unstinting in their praise for both our product and our enterprise. Another day when we went home extremely weary but satisfied and very happy. We'd finally made it.

WE'D BEEN FARMING seriously for about ten years and our credentials were pretty good. We were running about two hundred ewes over a hundred or so acres. Not a high stocking rate: as I noted earlier in this tale, the best guys can do three or even four ewes and their lambs to the acre, but with our fragmented holding and varied length agreements, we were doing OK. We used very little fertiliser and apart from a bit of topping and some rolling on the silage fields, our rented grass was pretty much maintenance-free. We were paying good if not exorbitant rents and had decent relationships with all our landlords. We even had two thirty-acre 'proper' farming sized blocks, one where we made good silage and the other providing much of our summer grazing. The latter made moving sheep so much easier either by just opening a gate or by using paths and tracks. We even had one delightful route down a tree covered lane through which the grandchildren loved helping us move sheep. In good years we made extra hay or silage on any surplus grass and sold it to our neighbours to help our winter cashflow.

We were selling about three hundred and fifty lambs a year, all through our retail business. Occasionally we'd need to buy in a few store lambs in September, but for the most part we were able to match our own lamb supply to the demand from the retail business. We'd get the best ones away in early July

'...a tree-covered lane through which the grandchildren loved helping us move sheep.'

and then manage the flow of lambs right the way through to June the following year. This took some doing as we've seen in earlier chapters but we'd become good at it and rarely ran out before the new season lambs (known affectionately round here as 'Springers') were ready.

We'd become expert at managing the feeding of our ewes and lambs to keep them healthy and productive and we were making good progress towards finishing most of our lambs

'We were selling about three hundred and fifty lambs a year'

without recourse to expensive concentrates. If we'd had more land of our own, I'd have experimented with crops such as kale and turnips to fatten lambs, but with our system I had to rely on making really good silage with lots of protein in it so we could keep them growing.

We really had come a long way from those early days when I'd sneaked out from working at home to cut hay with a borrowed mower.

Our buildings and facilities were as good as we could get them, too small as ever, but we'd learned how to make the most efficient use of the space we had. Our machinery was reasonably modern and reliable – so I felt less like the poor relation now and careful investment in kit like the dagging crate had made our lives so much easier.

We were also very proud of our welfare standards – something our customers were increasingly aware of. Our lambing afternoons helped, as we could show how careful we were with the new lambs and we would often get comments like 'You really care, don't you!' Chatting with customers at markets and festivals also gave us lots of opportunities to spread the word and we even had several vegetarians who bought our meat and burgers 'as they knew it had been treated well.' The evening talks that I gave during the winter were illustrated with lots of pictures which also helped get across the message that we always put our animals' welfare at the heart of everything we did.

One of the promises I'd made to myself was that we'd never put old ewes into the cull trade – the market for ewes and rams once they are too old for breeding. I knew they'd end up in the slaughterhouses in Birmingham where there was an increasing risk that they'd be subjected to the repulsive practice of non-stun slaughter. The Defra slaughter statistics for 2024 show an increasing level of non-stun slaughter with twenty-nine percent of all sheep killed in this way, up from twenty-three percent in 2022. As I see it, if you're going to end an animal's life, then you should do it with as little stress and pain as possible. I reckoned our animals had worked hard for us and we owed them a duty of care. I just didn't want to take this risk now associated with the cull trade.

And I'm pleased to say we never did. All our old or unproductive animals went to the abattoir in the same way as the lambs and we had them back as mutton. We were lucky that the time we chose to do this coincided with the so-called 'mutton renaissance' led by an enthusiast called Bob Kennard and his book called *Much Ado About Mutton* and we took full advantage of it. Over the years we built a good trade in mutton joints and diced meat which we sold for tagines and stews. Indeed, so successful were we that at times during the winter, our mutton sales outstripped our lamb. Our connections with the National Sheep Association helped here too as we were asked to provide mutton joints to roast on their stand at national events for samples that people could taste.

We had some great photo opportunities from this and some really useful publicity.

Our reputation as retailers and processors was every bit as high as our farming status. As I admitted earlier, I did it partly to be taken seriously as a business. I wanted us to do something different, something that had at least a chance of being financially successful. It was incredibly hard work and very demanding of time and resources, but we did it. We proved to ourselves and others that it was possible.

But most important of all, our move into processing had enabled us to add substantially to the value of our produce and to deal with the age-old problem of carcase balancing. Our burgers and lamb steak rolls had sufficient margin to enable us to trade at the bigger festivals and generate a significant increase in turnover. We really had moved the business up a gear. We'd finally achieved what we set out to do.

THE NICEST PART of this stage of our journey was the recognition we got from fellow farmers, neighbours and friends. I heard on the grapevine that a neighbour had said we were putting them all to shame. 'They're farming and paying rent which none of us could do and stay afloat,' was his assessment. They all owned their farms and paid no rent and although we could get into debates about return on invested capital, in their terms, we were doing better than them. I'd occasionally mused that if we'd not had rent to pay, we'd have made a nice margin from our business, albeit with a lot of unpaid family labour.

We also received some unexpected recognition from the local agricultural college which brought a couple of minibus loads of third-year students out to see us. I knew some of them – or rather I knew their fathers and it got back to us that they and the college had been very impressed with what they'd seen. It was particularly pleasing to chat with some of them who weren't from farming families. 'You've shown me that I could get into farming without a farm to inherit,' said one. 'It's clearly been hard graft but you've done it. It's inspiring!' I almost choked on my coffee. This was real recognition from the next generation.

And the accolades didn't stop there. You'll remember that we decided never to enter any more competitions after the Flavours of Herefordshire 'black tie' experience, a decision that we stuck to rigidly. But we did do well in two other competitions, ones that we didn't even know we'd entered!

The first was again at the Hereford Festival. We were winding down from another hectic show one wet Sunday afternoon, when one of the visiting celebrity chefs, Jean-Christophe Novelli, arrived at our pitch. I thought they were on the cadge for free samples for a cooking demonstration – 'We'll mention your name on the stage' was the usual offer – but he was there to award us with the prize for Best Stall at the festival. It was a lovely surprise and gave us more great photo opportunities and press coverage, but the best bit was his comment.

'You are doing great things here without lots of fuss and expensive promotion. Keep it up!'

'I like to see people who just quietly get on with producing good food,' he said. 'You are doing great things here without lots of fuss and expensive promotion. Keep it up!'

The second was probably our greatest achievement and certainly the most special to us. We had been nominated by farming colleagues for Sheep Farmer of the Year, an award sponsored and organised by the *Farmers' Guardian* newspaper. This was amazing recognition for us and a real acknowledgment of the respect with which we were held within the industry. We had to produce a formal statement of our business ethos and we had some professional photographs taken at Ludlow Farmers' Market. This was followed by a telephone interview with one of the paper's journalists who asked lots of questions about business direction and future plans. I'm not sure I answered these very well. We were so pleased to have reached a level where we weren't chasing the next sale and constantly trying to cut costs that I'd not given a lot of thought to continued growth.

We attended a glittering awards ceremony with Ross and Hannah and had a fantastic night, making memories among those whose opinions mattered to us. We didn't win, although

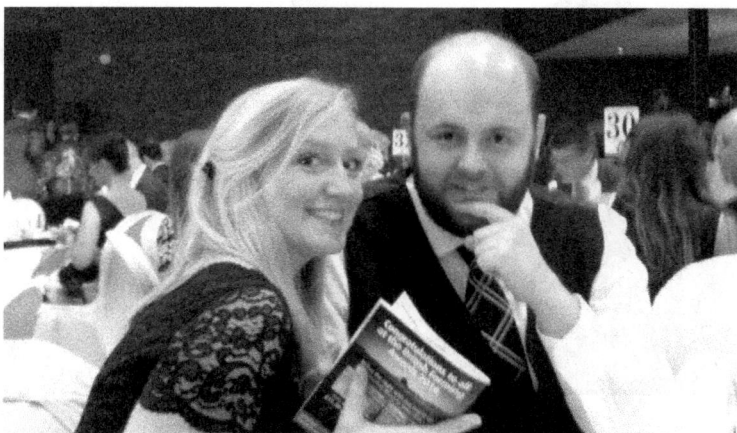

'Making memories among those whose opinions mattered to us'

'We were to feature on Malcolm Boyden's morning show.'

I did straighten my tie again, just in case, and it was wonderful to be part of such a prestigious event. This time we returned home with a warm glow, a real sense of pride and a feeling that all the effort we'd put in had been worthwhile.

We also became media stars in our own small way. I got a call from a researcher at BBC Hereford and Worcester, our local radio station. 'We're looking for a local food producer where we can follow the food from the field to actually eating it. We're going to call it *Field to Fork.*'

'That's interesting,' I responded carefully. With the vegan attacks still a fresh memory, we had become cautious about too much general publicity. 'OK, let's hear a bit more about it but in the meantime, how did you find us?'

'Oh we just Googled local food producers and your website came up!'

This was encouraging. It looked like the work we'd put into our web presence was paying off; we were getting known. And then the low blow, 'We asked at the abattoir at Leintwardine but they said they didn't know anyone.'

So much for over ten years loyalty. Every week without fail

'She's obviously read the book.'

we gave them our business but we were clearly not front of mind that day.

But this *Field to Fork* thing 'had legs', as we media types say. We were to feature on Malcolm Boyden's morning show and the project would include recorded interviews on the farm and some live phone-in discussions. Malcolm and his producer, Stewart, duly arrived one busy afternoon during lambing and interviewed me and Frances. We talked about the farm and what we did and Stewart took lots of pictures of baby lambs and pregnant ewes, but sadly none decided to give birth. 'That's OK,' said Stewart, 'we can choose a couple of lambs born today and follow those through the system.'

So we selected a pair of newly arrived twins and made a great show of recording their tag numbers so we'd know who they were as we and they progressed through the year. Hands were shaken and goodbyes said and as Malcolm was about to walk out through the barn door, I heard the tell-tale sign of a ewe starting labour.

We dashed back in and Stewart started recording my commentary and filming the event on his phone. It was a text book delivery. 'She's obviously read the book,' I said, much to Malcolm's amusement, as we watched the lambs arrive. Two beautiful lambs, both up and suckling within a few minutes, recorded live for transmission the following morning and filmed for their website.

The reaction we got was tremendous. It seemed to really strike a chord with the listeners. Stewart and Malcolm came back the day the lambs went outside with their mums and then at various points when we were weighing and treating them as the year progressed. We did several phone-ins with Malcolm and answered listener's questions about the farm and the animals, including a discussion about how sheep keep cool when it gets hot. (They don't – they have to find shade and it's one of the reasons that we shear them.)

As the time for their final trip approached, I had a chat with Doug the abattoir boss as I wanted this part to run smoothly. They were happy for us to record there but they didn't want to be included themselves. This wasn't unreasonable given the likely vegan reaction, so we decided to record a piece as we unloaded the lambs with me explaining that it wasn't my favourite job. I went on to say that the animals wouldn't have had the great life they'd had – indeed they wouldn't have been born at all – if we didn't do this. I was expecting a big uproar when this session went out, but we only had two negative reactions – one which I couldn't possibly print here and the other suggesting that I get another job.

For the final episode, Malcolm and Stewart came to the butchery one afternoon and filmed and recorded me as I cut up the lambs they'd followed through the year. I always found that a bit difficult, if I could actually identify the carcase with the live animal I'd seen running around, but we had a very calm and sensible discussion about respecting the animal whose life we had taken. We then took two lamb cannons into the kitchen and Frances cooked them for us to eat. Malcolm did a wonderful roundup of the year, effortlessly describing the cycle of the farm as only a pro broadcaster can and finished with two reflections. The first, that we'd never become millionaires by farming sheep, and secondly that our life and our work were intermingled to such an extent that we really couldn't stop! Both very prescient observations.

I'd had my doubts about this part of the series but it worked really well. We were praised for our sensitive handling of the issues and the reaction we got was entirely positive. Malcolm gave us lots of publicity for the next lambing afternoon sessions that were only a few weeks away by then and we were pleased to see lots of new faces who'd heard us on their radios – including someone who asked to see the ewe 'who had read the book'!

IT WAS HARD NOT to feel just a bit pleased with ourselves. There'd been plenty of people wanting to discourage us; the family thought we were mad, our friends couldn't understand why we worked so hard and even our farming neighbours I'm sure secretly wondered why we bothered. But we'd done it. We'd created a business that could make money, albeit on a small scale, and we'd done it from a standing start – no inherited land, no special circumstances, no 'silver spoon'.

But the challenge now was to work out where to go next.

9. WHERE NEXT?

I'D NEVER REALLY THOUGHT about the long term. All I'd wanted to do was grow the business. To create something that we could be proud of, something that would 'wash its face' financially and to achieve that elusive scale that would enable us to stand back a bit and employ some help. We'd managed the first two but the third was still some way off. We were still working seventy-hour weeks on our own, with virtually no time away from the farm. As a friend remarked, the more successful you are, the more your lifestyle deteriorates!

I was on my way to the abattoir one morning during lambing. It was a grim day with the rain coming down in sheets and I'd been up all night, so I maybe wasn't in the best mood for considering our future. I was thinking about how we could make things easier for ourselves and for the first time I began to wonder just how much longer we could keep doing this. I'd dismissed these ideas when others had raised them, mostly family and close friends who were worried about us, but was this really something we could sustain as we got older?

It was hard physical work, and the business was too big for the two of us to run comfortably but couldn't be expanded without spending a lot of money. We needed either to invest in more sheep and grazing land or fund a major expansion of the retailing operation with bought-in lamb. We didn't want to do the latter and, at our time of life, the financial risk of expanding the farming was maybe something to be wary of.

I put it to the back of my mind; there was no way I was giving all this up, we'd worked so hard to make it happen and it was my dream, after all. I'd occasionally wondered idly, when attending farm sales, what it would be like when it was our turn to receive the 'Thank you for your instruction' speech at the start of our own sale, but only in the abstract – there was no way we were going to stop just yet!

We carried on through the summer and autumn, with the silage making and the festivals, sorting the ewes, putting the tups out and all the usual routine and I forgot all about my 'wobble' earlier in the year.

But alongside these financial and growing-old pressures, there were other developments that we had to think about. The first was growing competition for the 'food to go' trade at festivals. Our offer was based on the local farmer bringing his product to a festival and selling to an interested and engaged public. Lamb burgers and lamb steak rolls were the mainstay of our income and they were popular with the more traditional festival-goer and with families. But we found ourselves competing with a whole range of new food offerings, brilliantly created by a new breed of professional vendors who were mostly trained chefs. These offerings included Mexican flavours, Piri Piri Chicken and a host of others, mainly aimed at the younger customers with greater buying power and more adventurous tastes. Our offer was becoming a little staid by comparison and although still popular, it was supported by a smaller and smaller proportion of the customers at each event. We'd need a radical rethink before long to keep this vital revenue source going. Such a change would involve more investment and moving even further away from our basic ethos as food producers.

There was also the increasingly vociferous vegan lobby that was damaging meat sales for all of us. There wasn't much we could do about it, other than to counter the absurd claims being made with level-headed factual information, but it was hard work. We took every opportunity to explain how meat production is part of a natural cycle and that removing animals from farming systems wouldn't work. We explained how much time and effort we put into our animal's welfare and while some listened, most didn't. The overall effect of the vegan lobby was – and still is – tiny in business terms, but the energy and time they absorbed was out of all proportion and was very wearying for those of us just trying to do a decent job.

Not as vitriolic or extreme as the vegans but enthusiastically supported by them (and equally tedious) was the endless promotion of plant-based diets. Variously justified as being better for our health and/or saving the planet from climate change, neither with much evidence, these ideas were just emerging as we were contemplating our future. The reasoning went something like:

Here's a crisis – Climate Change – we must do something.
There's something – Plant-based Diets – let's do that.

Ideas like this have flourished through lazy, uncritical journalism, sensationalists in search of click-bait, huge vested interests from religious groups and the food industry, and others. Just recently I heard a well-known TV personality talking about

'Dealing with climate change by eating less meat,' as if it were mainstream thinking and hard fact – which it isn't. Debunking these ideas is way beyond the scope of this book or my intellect, but I can recommend Jayne Buxton's *The Great Plant Based Con* which, despite its title, is an even-handed and thorough analysis of the scant, conflicting and highly biased 'research' evidence put forward for the benefits of plant-based eating.

The other challenge was the increasing need to farm sustainably. I mentioned regenerative agriculture or 'regen ag' earlier, but that's only one of many new ideas currently in vogue. Others include silvopasture and agroforestry which both involve mixing trees with conventional farming. I've spent much of my life introducing change into farming and I'm very supportive of new thinking, tempered only with the need for proper scientific evaluation to ensure the benefits are repeatable, genuine and cost effective. Sadly, these fundamentally good ideas, which are mostly about improving the soil, are being devalued by having labels attached to them. Most 'regen' practices involve increasing soil organic matter and hence water-holding capacity, nutrient content and, most important, carbon capture and storage. But it's all getting lost in the rhetoric. A friend once told me that if you preach a doctrine, you'll attract disciples but lose everyone else. The denigration of regenerative principles as 'hippy farming' by some of my farming friends seems rather to support that view.

But the notion of sustainable farming had also become embedded in the minds of many of our customers and we needed to take account of this in our marketing. The difficulty was that we didn't have a farm of our own. As described throughout this book, apart from a few acres at home, we rented all our grazing and, other than day-to-day management, we had little or no control over what could be done with it. Our landlords would certainly not have let us plant trees or multicrop their fields and some of the more extreme 'regen' practices like leaving long rest periods between mob grazings would have had us turned off the land as it became untidy and full of weeds. These practices also involved very low stocking rates and since we were paying by the acre, we needed to maximise our returns at that level. If we'd had our own land, I'd certainly have tried some of the new ideas but in these circumstances it just wasn't possible.

None of these challenges were at all unreasonable for any business but perhaps ones we needed to bear in mind as we contemplated our future.

OUR LAST LUDLOW FESTIVAL each year was always the Medieval Winter Fayre, a great money-earner but very difficult when the weather was bad. And this year was terrible, with driving rain, wind gusting to sixty miles an hour, and bedraggled customers paddling around in the mud. Frances arrived late on Saturday, having traded at a market in the morning, and I watched her struggle with a trolley across the mud, in the driving rain, bringing in extra stock. And at that point I decided. Enough was enough. We had to stop. It just wasn't fair on her or others who were trying to help us.

I mulled it over until the Sunday afternoon and as we stood there in our sodden gazebo watching the rain drive the last of our customers away, I suggested we might call it a day. To her eternal credit, Frances was very cautious. 'Let's think about it over Christmas and then see,' she said, although I'm sure she secretly wanted to jump and shout with joy and relief!

We discussed it over the holidays and decided that we would stop but that it would be kept confidential for a while. We needed a plan so that we could stop the farming over a sensible time period and we might even be able to sell the retail business and preserve what we'd achieved. I didn't want our various landlords to get to hear that we were stopping until we were on the grass for the last season, in case they decided to let it to someone else. Similarly, I didn't want our customers to go elsewhere, particularly if we planned to sell the business. The other risk was that the festival organisers would hear we were packing up and let our pitches to other traders – such is the competitive nature of these events.

Farming is an intensely seasonal business and exiting at the wrong time can lose you a lot of money. It was too late to stop that year, but we would make the following spring our last lambing. We would continue to trade through to December that year, put the tups out as normal in October and sell the ewes, in lamb (pregnant) in the following February. This would allow us to use up silage made during the year and continue to manage the grassland properly, prior to vacating it all in December. So we had one more 'normal' farming year to enjoy and plenty of time to plan before we started to wind the business down.

SELLING UP THE FARM was a carefully organised campaign, planned with the help of Michael, our local auctioneer. We spent a whole evening travelling round our groups of sheep with him helping me decide which sale to put them in. I am immensely grateful to him as without his help we would have struggled. We had various groups of lambs and ewes to go as well as the equipment and machinery – all to be prepared for specific sales to achieve the best possible prices. Ram lambs were to be fattened to go at the end of August to catch the special festival trade, ewe lambs for breeding and a fine batch of yearling ewes were to go in September along with some small store lambs which we didn't need. The rest of the meat lambs would go once we knew how many we'd need to continue trading to Christmas.

In parallel with packing up the farm, we quietly put the

retail business on the market. I approached several market traders, whom I thought might like to take on an additional source of income, including one who already farmed sheep and another who had all the facilities to process and pack the meat. Alas, both came to nothing

But then we had a bit of luck. A local farmer's son called Ed rang to see if I could recommend a source of insulated boxes to ship some of his lamb to friends in London. 'I'm not going to compete with you,' he assured me. 'We're only doing this for friends at the moment.'

'I can do better than that,' I said. 'The business is for sale if you're interested.'

He was interested and within a few days Ed and his partner Sheena were sitting at our kitchen table looking at figures and discussing how they might take it on. Ed was easy to deal with; he'd spent time away from the farm and knew about business and contractual matters. He and I drew up a contract of sale which included the use of our butchery and cold room for up to twelve months while he established these facilities on his own farm just up the road. We also included some of my time to help him get started and to teach him the butchery skills he needed.

THE LIVESTOCK SALES came and went and we did very well at most of them. We had a large batch of bigger lambs that I had intended to sell as stores but they grew so well that we sold them fat and topped the market that day in Ludlow. This was the most wonderful feeling as we rarely exposed our stock to the rigorous judgement of the market. It was a nice endorsement of our animals.

Unfortunately, our plans to sell the ewes the following spring proved rather more chaotic. It was now early September and having thought about it, my idea to sell them in lamb did

'It was a nice endorsement of our animals.'

carry some risk; they probably wouldn't look as good as they did now, having had a great summer on good grass, and their value would vary depending on how many lambs they were carrying. Twin-carrying ewes would be worth more than now as would singles but triplets and empty or barren ones, much less. It was all just a bit too uncertain and after a brief chat with the auctioneer, I decided not to take that risk. We were booked into the breeding ewe sale two weeks hence and to be fair to Michael, he had suggested this as a less risky course of action when we were planning our sales back in the spring. He never said it, but I'm sure he thought, 'I told you so'.

With this huge weight off my mind, I decided to attend the sale the week before ours, just to get a feel for how it would work. This was a big mistake. The trade had dropped alarmingly and I saw decent quality sheep punished with rock bottom prices or even no sale at all. We'd spent many years building this flock and I could see it all being given away on a poor trade. This prompted yet another change of plan.

Although the trade was down, the best-quality sheep were still selling well so I decided to split the ewes into the really good ones and the rest. John, who was about to take over our land and buildings, was looking for some decent sensibly priced sheep and he took the majority of the older ewes and the tups at a very fair price. We then booked Charlie with his lorry to take the best quality ones into Ludlow for the sale.

I remember that morning well. I had sorted the ewes into groups the night before and carefully put them into pens so they could be loaded onto the lorry separately. At least their last night with us was comfortable in nice clean well-strawed pens. Sale day was grey and cold – we went out in the dark to check everything was OK and Charlie duly arrived at 6.30am to collect them. He had a different lorry this morning with 'Charles Rae' emblazoned on the door. 'The world knows you as Charlie,' I laughed, 'what's with the Charles?' It turned out that the previous owner's name had contained enough letters to spell Charles but not Charlie and he'd carefully unpicked the lettering from the door and rearranged it to save the cost of a re-brand. A little episode that made me smile, on an otherwise rather sad morning.

I followed him up to Ludlow, supervised the unloading – and then waited. 'Hell, they're some good-looking sheep – are they yours?' exclaimed a farming neighbour. 'They'll do well today, there's plenty of folk who'll pay for good ewes.' This was encouraging and lifted my spirits a bit.

Sue, my friend who used to lend us her tup, Basil, also came to offer some support, which was kind of her, and several other local farmers congratulated me on the quality of the stock. All this was reassuring and good to hear on a difficult day.

At last the sale started and I went to sit and watch the bidding. Again, they weren't selling that well and some very nice-looking ewes went cheaply. As it got nearer to my lot numbers, I made my way nervously to the little door behind the auctioneer's rostrum and watched the drovers bring my sheep up to the sale ring.

And then it was our turn. I slipped through the door and stood facing a room full of buyers with Frances sitting in the middle of them looking as worried as I felt – needlessly, as it turned out. The bidding went up and up and the youngest ewes went for more than I'd hoped. The next pen was the same. And the next. Then we got to the three year olds and again the prices were good. It was the same person bidding – he bought most of our sheep and even had to battle for the last pen as someone else wanted them too!

The final result was better than I'd hoped; we'd done well in a cautious market and I was relieved. I found Charlie and told him I wouldn't need him to take any of them home again and we then took Sue for a congratulatory bacon roll and a coffee. While we were eating, the buyer of our ewes came over to say hello and I gave him a 'bit o' luck.' This is an old custom which has almost died out everywhere else but still happens in these parts – when the seller gives the buyer some cash to wish him 'luck' with the animals he's bought. In my case, £10 luck money for nearly £7,000 worth of sheep didn't seem a bad deal.

I BECAME A BIT OBSESSED with 'lasts' during 2017 – our last lambing, our last-ever lamb born, our last silage season – I have hundreds of photos and videos with which to bore the family for years to come. But perhaps the most memorable 'last' was our final food festival.

The Shrewsbury Winter Festival was always a good one for us. Organised by our friend Beth, it had a nice atmosphere as it was close to Christmas and we liked it as it was inside a huge marquee and therefore fairly weatherproof. We had a good Saturday and sold a lot of burgers and meat and were on target for

a good if not great last-ever appearance, but there was snow forecast for the Sunday. Many of the traders packed up on the Saturday evening, but we couldn't get our chiller and cooking facilities off the site until the next day, so Ellie and I cleared up for the night and went home.

At five the next morning, I looked nervously out of our bedroom window to find several inches of snow on the ground and still more coming down. Frances and I left in our Land Rover at six, planning to trade for the day and bring as much home as we could that night. We always cleared Shrewsbury on the Monday morning as, a bit like Ludlow, it was much easier to dismantle everything once most traders had left but it looked as though today would be different.

The journey took forever, mainly because of other very timid drivers on the road, but we eventually got there to find that the snow had caused some of the marquee roof to collapse, luckily some way away from our stall. We all looked at each other glumly. This wasn't going to be a great day – even if our customers could get onto the site and it would be something of a damp squib ending to our festival career. But

Beth and her team pulled out all the stops and got local radio and TV involved, offering free entrance to those who could walk there.

We sold a little, mostly to those sympathetic to our 'bravery' for being there at all, and barely covered our costs. By three o'clock we were all getting worried as it had been snowing all day. It was clear that even if we got home that night, we'd never make it back again in the morning. So we dismantled everything and stuffed it all into the back of the Land Rover. Poor Frances was wedged in the passenger seat with boxes and bags all around her and that which we couldn't get in, we just abandoned. Then, after much hugging and kissing of Beth and her team, we left to brave the journey home.

It was a nightmare, even worse than the trip up, although at least the timid drivers had all gone home or given up. The road was deserted as we trundled slowly home through the blinding snow. I've never driven in a 'white-out' before or since, but I now know how disorienting it can be. We found ourselves planning the best route in to the farm from the main road, avoiding the steepest hills and deciding which likely trouble spot was closest to walk from if we did get stuck. We just made it although we were the last vehicle for several days to battle our way through the foot-deep and drifting snow in our lane.

It was a truly memorable end to our festival career and a tale we would dine out on for years afterwards.

WE'D MADE A GOOD job of selling the meat lambs and would have none left when we ceased trading at Christmas. Unfortunately we weren't quite so well organised with the mutton ewes and, despite offering huge amounts of free meat to our friends and regular customers, we were left with four of them with no home to go to.

I asked around our friends to see if anyone wanted them as

'We were left with four ewes with no home to go to.'

pets or lawn mowers and eventually sold them to a fellow market trader who wanted to give them a decent 'retirement'. Ironically, she was a vegan and given all my reservations – some might even say prejudice – about them, had arranged for them to go to a farm animal sanctuary. I was uncomfortable about this; these places were sanctuaries for ill-treated or abandoned animals. My animals were neither. I was just trying to keep them out of the cull trade and I should have tried harder, either to sell them as meat or re-home them properly. The resources of this place were being used, inappropriately in my view, to help me salve my conscience. I resolved it, for myself at least, by giving them a lorry-load of silage I had left over. It was probably enough to feed those four sheep for the rest of their days and the sanctuary owner was delighted, so we ended up with a happy outcome for both the sheep and me.

Once Christmas was out of the way, I then had until 24 February to prepare the equipment (what farmers call deadstock) for a sale so that it coincided with buyers thinking about lambing and their need for sheep equipment. We decided not to sell the tractor as I still needed it and we kept back some of the machinery to sell at a later date.

I spent much of January and early February in the barn, cleaning and repairing all the kit – I even painted a hundred or more sheep hurdles with red oxide paint. I replaced the decks in the sheep trailer and polished it until it shone as this was to be the main 'draw' for the sale. On the Thursday of sale week, Ed helped me move all the kit to the auctioneer's sale field – it took most of a day with two tractors and trailers – and I then spent hours arranging it carefully to show it off to best effect. The auctioneer helped me organise the more expensive items and then booked it all in ready for the sale the next day. 'Don't get here too early,' was his parting shot, 'we know what we're doing. You just enjoy the day!'

It was a strange feeling walking the dog the next morning; this was the end of our business, the sheep had long gone, but this was the final step. After today we would no longer be farmers.

We arrived on the sale field, way too early, to find everything

'Don't get here too early, we know what we're doing.'

in order. The auctioneers did indeed know what they were do-ing and they'd advertised the sale very well. Frances and I po-sitioned ourselves where we could talk to people about the kit and also to keep an eye on some of the more 'portable' stuff like the tag reader. In our early days of attending farm equip-ment sales, we'd bought some feed troughs and on returning with the car to collect them had found them gone – alas not everyone attending these sales is honest and supportive – al-though of course most are.

We had a lot of people turn out for us on what was a bit-terly cold day and we had a great sale. One of my more ratio-nal friends, standing behind me as some gates went under the hammer muttered, 'This is crazy: I can buy those new for less than that!'

'Shhh,' I replied with a scowl. This often happens at farm sales; bidders get carried away or maybe they just like to sup-port their fellow farmers as they pack up? In any case, my hard work painting all those hurdles paid off as we achieved an av-erage of new price across all of them.

The fertiliser spreader sold for about twice what I'd paid for it, having been cleaned up and a broken part repaired, and then the group of bidders and on-lookers reached the final lot – our stock trailer. I thought it might fetch £1,200 if we were lucky. It was 'tidy' which is Herefordshire for 'well looked after' but it was an old model. The bidding shot straight up to £1,000, then £1,500 and eventually settled at £2,400 – a fantastic end to a sale where we realised about twice the total value that we expected and a real vindication of the work we put in to get everything looking as good as we could on sale day.

As we left the sale field munching the obligatory bacon but-ty (I still go to the odd farm sale for those alone), Michael came over to shake our hands and remarked, 'The only real mistake you made was not to include the tractor. You've got a good

'I drove it sadly to the sale ground and then cleaned it again.'

reputation here now and you should trade on that before they forget you.'

I took his advice, rang him on the Monday morning and booked my beloved expensive tractor in for the next sale. I touched up the odd scratch, cleaned the windows until they looked like new and polished it to within an inch of its life. I even used lacquer spray to clean the dashboard and controls, and I have to say it looked rather good. I drove it sadly to the sale ground and then cleaned it again, even removing the mud from the tyres.

The 'tyre kickers' were out in force the next morning. 'Ah this is the older model with the smaller engine,' said one,

'They're not worth as much, you know.' Others queried its hydraulic capacity although most conceded that it was 'tidy'. It was another wet and cold day as the auctioneer and his following reached me and I started the engine and moved the loader up and down in a desperate attempt to convince them to bid.

At that moment, a pleasant young chap climbed the steps to the cab and said, 'Is she yours? Does everything work and is she a good yard tractor?' to which I replied, 'Yes, yes and yes.' The bidding started way below the reserve and faltered at £20,000. The auctioneer looked nervously at me and I shook my head. I wasn't going to give it away, but then my new friend jumped down and pushed the bidding up and bought it for just over the reserve. It's a strange feeling sitting in a machine that's being auctioned – I could see most of the bidders but not everyone, and the inscrutable looks on the tyre-kickers' faces were very disconcerting. Anyway, the result was that I sold it for what I paid for it as the dealer had promised, of course, quietly forgetting about the ludicrously expensive repair!

At this point we went away on holiday – our first proper break for twelve years. It was lambing time and Frances had booked a week in Cornwall, well away from sheep so we could enjoy our first spring together for some time. She had actually booked it when we first decided to pack up and I'm pretty sure it was at least partly to make sure I didn't change my mind.

THE FINAL PART OF the jigsaw was to get Ed and Sheena up and running with the retail business.

Handing over the business on 1 January was ideal timing as it's very quiet at that time of year after the demands and stresses of Christmas. My friend Paul, the stockman at the abattoir, helped Ed get organised with his first batch of lambs and he returned with them to put in the cold room a few days later. I then spent a day with him, teaching him how to cut up the carcases for markets and how to make burgers. He learned very quickly and within a couple of sessions, he'd mastered the butchery skills he needed.

We went together to his first farmers' market in Ludlow where he quickly got to know his fellow traders and proved to be a great salesman for his products. I was quite sad to end that session as it would be my last-ever market day, but it was good to see him make such a great start.

I helped him set up at the Ludlow Spring Festival in May, introduced him to the festival team and worked on the stall myself for the first day. It all went very well and he sold more than planned. I even helped with some more prep on the Saturday night to replenish their stocks for the next day. But that was it: I was finished and as I said goodnight, with only a slight twinge of sadness, I knew it was now up to them.

They had a great first year, sales were pretty much on target and they seemed to be enjoying themselves. They re-branded the business, changed the name and generally made it their own. The last part of the transfer was to move the butchery and cold room to his farm which eventually happened the following year. Whyle House Lamb was no more and Perrywood Farm was up and running.

We were delighted to see our business continue. We had worked so hard to build it up and it was nice to see it flourishing under Ed and Sheena's management. From a selfish point of view, it was also very rewarding to give an enthusiastic young couple a start in what we hoped would be a successful new venture for them.

As for us? Well, we just had to get used to not being farmers any more.

10. Back to a Normal Life

A CHRISTMAS DAY WITHOUT the sheep to feed was unusual for me. I confess I used to quite enjoy the routine. A chance to escape the mayhem in the morning for a couple of hours. I'd feed the ewes in the barn and then take a leisurely drive around the sheep that were still out in the fields. We'd put everything out ready on Christmas eve so it was just a matter of filling the feeders and checking that the animals were OK. But I used to take my time, especially if it was a nice day. I'd often see friends and neighbours out for a walk, all sporting ridiculous Christmas jumpers, and I couldn't help feeling a bit superior. They were being frivolous whereas I was out earning a living and caring for my animals!

I then did it all again in the afternoon. A chance to get some fresh air while others slept off the excesses of lunch. I didn't do the outside groups this time but I used to spend an inordinate amount of time feeding and checking the ewes in the barn – often with a grandson or two in tow. It was a lovely time, the rustle of the straw as the sheep moved around and the warm glow from the barn lights as dusk fell outside. The boys would help me measure and put out the feed and the sound of a couple of hundred sheep munching contentedly was wonderfully reassuring.

But this year I didn't have that excuse and I had to spend the whole day with the family!

A COUPLE OF DAYS later we attended a friend's lunch party. 'If anyone else jokes about us being retired, I'm going to thump them,' I said as we walked up to their door.

'Hi, so this is what retirement looks like,' laughed our host.

'I'll hold your coat,' murmured Frances.

I didn't particularly enjoy that party. I was feeling lost and bereft and everyone seemed to be happy that we'd packed up. 'Oh it must be so nice not to have all that worry,' said one neighbour. 'And you don't have to get up so early now!' All completely missing the point. Later in the afternoon the local retired GP's wife came up to me.

'Are you OK Andy?' she said.

'Yeah, I'm fine,' I lied.

'No really, are you all right? You seem a bit low.'

We had a long chat and she told me how her husband had had to deal with retirement – suddenly not being the local GP any more had been quite a change for him. That was exactly how I was feeling. I'd worked so hard to be part of the farming community, to be accepted as one of them. Respected even. And now it was all gone.

It wasn't all gone, of course: they were still my friends, I could still stop for a chat in the lane, but it was different now. All my life, all I'd ever wanted was to be a farmer and, against all the odds, I'd succeeded. And now I'd dismantled it all and walked away. The hardest part was that friends and most of the family didn't understand. They were relieved on our behalf that we didn't have to work so hard, which was nice of them, but mostly the verdict was that, 'At last we'd seen sense.'

I found this dismissal of our achievements hard to handle for a while. There was no malice involved and it was meant with the best of intentions, but it rankled. On bad days I would anguish, selfishly, over their insensitivity, 'Couldn't they see how much it meant to me? How I was feeling?' And of course they couldn't. It wasn't their fault. They hadn't been associated with farming all their lives and they didn't appreciate the way it becomes part of your life, part of your existence. It's not just a job.

But of course there was someone else involved in all this. Frances was clearly happy that we'd finished. She'd been amazingly supportive, but it wasn't her dream. It wasn't her lifelong ambition to farm and, having grown up on the Wirral in Cheshire, she'd had very little involvement in country matters, let alone practical farming, until we met. I knew she'd had no idea what she was letting herself in for when we started and I'd only introduced the downs of farming life to her as they happened. I was feeling sorry for myself but she'd moved on and, not unreasonably, was enjoying a normal life again.

So I decided to turn to my farming friends who at least understood. Their views were more mixed, some wondering why we'd given up what we'd clearly worked so hard to achieve and others apparently envious that we'd 'bitten the bullet' as they put it and stopped before we got too old. The one common theme though was how they admired what we'd achieved. And that made me stop and think.

We'd bought Whyle House with a hefty mortgage and lots of enthusiasm. It was to be a hobby, nothing more. We'd funded our early farming journey from very limited savings and our full-time income. We'd slowly increased our sheep numbers from ten to twenty to thirty and then the first big leap with the sixty or so from Mr Severnoaks. We'd struggled with tiny pieces of rented land, trying to gain sufficient credibility to get onto the rented grass network. We'd dealt with eight different landlords, gradually increasing the size and quality of our land holding but continually struggling with the logistics of having sheep spread across the county. Our equipment and machinery fleet had migrated from home-made, worn-out junk to serviceable and reasonably modern kit, all of which was paid for.

Looking back, it had been an incredible feat to get the farm to this size and scale from a standing start and with no external help – apart from an accommodating bank manager.

On the retail side, our progress had been similarly impressive. Selling fourteen lambs to family and friends in our first year and painstakingly building our reputation from there. Learning to butcher, learning to sell and later making the substantial move into processing. Growing the farmers' market trade from one to thirteen a month and the lucrative but wearying festival outlets to nine annually. Dealing with demanding customers in both retail and wholesale, protecting our reputation and managing the relationships necessary to keep it all running smoothly.

Put like that, we'd a lot to be proud of. The transition out of farming had been hard but we could be very pleased with ourselves. I was beginning to feel better about leaving the industry I'd grown up in and loved all my life.

As time moved on, I got busy with other things. I helped John occasionally with his sheep and I bought another tractor (a much older and cheaper one) so I could keep our ground tidy and do a bit of contracting. I also took on some mentoring work as part of a Prince's Trust Funded Project which got me out onto local farms, helping the next generation to take over the reins from their parents.

Frances meanwhile, relieved of her farming commitments, was able to pursue her own interests. Her new life included volunteering at the local tourist office and driving for a hospital car service known locally as Community Wheels.

Despite some tricky moments, our relationship had survived working together almost continuously. Personally I think that's more of a tribute to her than me. We'd had one near tragedy early in our farming career which really tested her commitment, way beyond what was reasonable.

'Fancy helping me with the wire between Leasows and Shop Field,' I mentioned casually one Sunday lunch time. 'I've done the posts, which is the hard part, and putting up the wire is so much easier with two. We can get it finished today.' Frances, well accustomed to my, 'It won't take a minute' jobs, sighed and followed me out. It was a lovely autumn afternoon, still warm enough to work without a coat and dry underfoot. We rolled out the wire netting in the warm sunshine and fixed it carefully to the post at one end. I then gently pulled it with the tractor from the other end. Local fencing lore dictates that you do this until it stands upright. At this point, according to the experts, it will be at the right tension.

We worked our way along the fence stapling the wire to the posts, 'Every other horizontal strand and not too tight else you'll break the wire.' A little snippet of farming wisdom I remembered from Horace, my work mate at Clarke's.

We completed the netting by fastening it to the heavy gate post at the near end and released the tension by loosening the tractor handbrake. The posts groaned and creaked slightly but the fence stayed up, perfectly tensioned.

'That's not bad,' I ventured, 'we're getting the hang of this!'

Frances smiled, 'It does look pretty good, I have to agree.'

'Just the top strand of barbed now and we'll be done.'

The process was the same as with the netting. Having fixed the barbed wire at the far end, we put a rod through the reel and, each taking a side, unrolled it as we walked back to the gate.

'I'll just pull it tight,' I said as I clamped it to the front of the tractor. 'If you go and get the staples and hammer, I'll join you.' She set off across the field as I started the tractor and engaged reverse. I eased the clutch and the wire tightened. 'Just another inch,' I reckoned. And then it happened. A rifle 'crack' split the air as the wire broke and whipped across the field, wrapping itself around Frances as it went, pulling her to the ground.

For a moment I thought I'd killed her. I jumped from the tractor and raced across the field to find her tangled in the wire with lacerations to her face and neck. I grabbed the fencing pliers and cut the wire off, ripping my own hands in the process. She was badly cut but she was OK and mercifully the barbs had missed her eyes. I ran back to the house for the car and took her to hospital. The nurse didn't quite get what had happened and when we explained about an altercation with a barbed wire fence, her incredulous response was, 'Well didn't you see it?'

It was some weeks before she fully recovered and twenty-five years later, she still has scars to tell the tale and I still have enormous guilt. It could have been so much worse. She had every right to walk away at that point. As far as I'm aware, 'in sickness and in health' doesn't cover being wrapped in barbed wire. But, to her eternal credit she didn't. Her only stipulation was that she never came fencing with me again!

WHILE WORKING – what friends called doing our proper jobs – we'd got used to going our separate ways every morning and suddenly we were with each other all day, every day.

Frances had made the painful (literally and metaphorically) transition from interested observer to skilled and active business partner with weary stoicism and (mostly) good humour. We both had to learn about retailing from scratch and I think that brought us closer together. She understood the end product and I learned how to prepare it and then we both learned how to sell it. I think her greatest contribution, apart from her financial acumen, having been in business before, was her ability to keep going. Her tenacity, even during the worst weather or the most disastrous setbacks was an inspiration to me.

So, despite some difficult times, I like to think we managed our farming lives and the move back to a normal life with reasonable grace and mutual respect. We rose to the challenges, we supported each other and we took genuine pleasure in each other's achievements. We're now 'retired farmers' rather than 'ex-farmers'. I was sad to see my dream dismantled but reassured to see the business continue and Frances was quietly relieved for herself and me. We'd navigated our way back to a normal life.

SO THAT'S THE END of our tale. From a simple hobby to a busy and extensive farming and food retailing business. It never got to be big: it was never going to with our limited resources. If we'd started with a plan and some idea of the scale we were aiming for, we could have done things differently. But we had no idea what the opportunities were or how the retailing would develop. Trying to convert a hobby into a business gave us unnecessary challenges but made the early years lots of fun. We made some silly decisions but we got a lot right too.

If I did it again, I'd aim higher and I'd invest first in the things that earn money. Buildings and equipment are nice to have but they don't deliver income. Sheep and lambs do. And when I did need equipment, I'd buy it sensibly and not waste time trying to save money by making it myself. I'd also have got as big as we could, as quickly as we could, to generate enough cash to keep growing.

One of the nicest things about this journey has been the people. The farming community is a lovely thing to be part of and our venture into retailing introduced us to some fascinating and very helpful individuals. Everyone was welcoming and supportive – not always attributes of business people in other fields.

So that was it, we stopped with mixed feelings. We didn't miss the mud and the cold, the pressure of preparing for the big festivals, the very early mornings and late nights (and even no nights) and the relentless routine of the retailing. We did and still do miss bringing new life into the world, the challenges of running a farm, being part of a very encouraging farming community, and above all, those lovely people we met on our journey.

I've reflected many times on our story and the hard-won knowledge we gained along the way. Perhaps others will learn from our mistakes, perhaps not. But the most difficult-to-answer question is 'why?' Why did we do it? Why did we put ourselves through the pain and penury?

Maybe we'll never really be able to answer that one, but given the chance, would we do it all again? Of course we would!

ACKNOWLEDGEMENTS

WHEN I STARTED thinking about who to thank for their contributions to our story, the first name that came to mind was Geri, our sheepdog. A little unusual perhaps but my first acknowledgement must be to him – my loyal workmate and pal for all those years. Even now I get a lump in my throat when I think about him because, in truth, we couldn't have done this without him.

Of course many people helped us in many ways. Thanks especially go to our farming friends, and neighbours who were unstinting in their support and encouragement – particularly Cecil Price and Tim Brookes who mentored me and encouraged me to keep going, even in the worst times. Adrian the shearer, John the sheep scanner, Harriet the vet, Kevin the silage contractor and Kevin the groundworks guy were all a delight to work with and made the job so much easier. Our landlords were patient, supportive and encouraging, several saying that their land never looked as good as when we farmed it, which was nice to hear. Joe our farm helper and Tomos helped in both the lambing shed and the butchery and were great fun to work with and see on their way into adulthood. Thanks to the many lambing students who helped us and made us smile even on the worst of days, including Ellie and Abbie – Abbie for having sufficient enthusiasm to come twice and Ellie for all her help both with the sheep and later at the festivals. It was wonderful to see these young women make their way very successfully in the professional world, and to remain in touch with them.

On the retail side of the business, Barry's unfailing good nature in the butchery got us through many a tough day, and my good friend Paul, the stockman at the abattoir, made those difficult trips just that bit easier. Our fellow traders on the markets and at the festivals, including David and Janet Griffiths, Nick and Faith Wenden, Chris Davies, and Chris and Guy Tudge became very good friends and shared the ups and downs of retailing life with humour and tenacity.

And of course, none of this would have happened without our customers. Supportive, encouraging, and cheerful, they made the long days on the markets and at the festivals worthwhile. Many became good friends and most stayed with the business to continue to buy from Ed.

Ed and Sheena who took on the retailing business and John who bought many of the sheep and went on to rent our land and buildings – all three of them helped us through the difficult period of packing up and gave us the reassurance that all our hard work had created something that could continue.

Our friend Michael Thomas, the auctioneer, was a source of advice and help over the years, buying sheep for me in the early days and helping with the difficult and ultimately unsuccessful attempt to take on Bert's land. He also provided us with tremendous help when we were packing up and made a difficult eighteen months much easier – and delivered a great outcome for us.

This book wouldn't have got off the ground without the help and support of Craig Hillsley whose editing made the thing readable, and publisher Chuck Grieve of Mosaïque Press whose creative partnership enabled me to produce this wonderful book. Thanks also to friend and former colleague Mel Lawrie who did a sterling job proofreading and also kept the agricultural references and terminology accurate and truthful. Heartfelt thanks also to the Kingsland writing group and our leader, Fay Wentworth, for their support, encouragement and belief that I could actually do this!

And finally, our family – children and grandchildren – all did their bit. At festivals, in the butchery and of course in the lambing shed, they all worked so hard for us.

One last thank you – a personal one from me to Frances, my lovely and long-suffering wife. Those hours spent packing meat and making burgers and freezing with cold on market stalls; the 5:30am starts, the early shifts in the lambing shed, the tireless efforts to rear tiddler lambs and helping vaccinate sheep for scab in the snow. Supportive, even on the worst of days, and good humoured, even when we were involved in that notorious relationship-breaker – moving sheep.

This was my gig: it was my dream and you helped me make it happen. Thank you.

www.ingramcontent.com/pod-product-compliance
Lightning Source LLC
Chambersburg PA
CBHW071419090426
42737CB00011B/1509